Machine Learning for Semiconductor Materials

Machine Learning for Semiconductor Materials studies recent techniques and methods of machine learning to mitigate the use of technology computer-aided design (TCAD). It provides various algorithms of machine learning, such as regression, decision tree, support vector machine, *K*-means clustering and so forth. This book also highlights semiconductor materials and their uses in multi-gate devices and the analog and radio-frequency (RF) behaviours of semiconductor devices with different materials.

Features:

- Focuses on semiconductor materials and the use of machine learning to facilitate understanding and decision-making.
- Covers RF and noise analysis to formulate the frequency behaviour of semiconductor devices at high frequency.
- Explores pertinent biomolecule detection methods.
- Reviews recent methods in the field of machine learning for semiconductor materials with real-life applications.
- Examines the limitations of existing semiconductor materials and steps to overcome the limitations of existing TCAD software.

This book is aimed at researchers and graduate students in semiconductor materials, machine learning and electrical engineering.

Neeraj Gupta is an Associate Professor at Amity University Haryana with over 16 years of teaching experience. His expertise includes VLSI design, low-power and analog design, AI and embedded systems. He has published 40+ papers, two book chapters, one book and 12 patents, and has received the Best Researcher and Best Teacher Award (2024).

Rashmi Gupta is an Assistant Professor at Amity University Haryana with 13+ years of experience. Her research interests include AI, software engineering and IoT. She has authored 20+ papers, two book chapters, one book and five patents.

Rekha Yadav is an Assistant Professor at DCRUST, Murthal. She specializes in semiconductor device modeling and VLSI design with 15 years of experience, over 30 publications and four book chapters.

Sandeep Dhariwal is an Associate Professor at Alliance University, Bengaluru. With 14+ years of experience, he focuses on low-power CMOS and semiconductor modeling. He has published 40+ articles, three books and holds three patents.

Rajkumar Sarma is a Postdoctoral Researcher at the University of Limerick, Ireland. With 11+ years of experience, his research spans digital VLSI, FPGA prototyping and quantum architectures. He has 25+ publications, 15+ patents and two books.

Emerging Materials and Technologies

Series Editor: Boris I. Kharissov

The *Emerging Materials and Technologies* series is devoted to highlighting publications centered on emerging advanced materials and novel technologies. Attention is paid to those newly discovered or applied materials with potential to solve pressing societal problems and improve quality of life, corresponding to environmental protection, medicine, communications, energy, transportation, advanced manufacturing, and related areas.

The series takes into account that, under present strong demands for energy, material, and cost savings, as well as heavy contamination problems and worldwide pandemic conditions, the area of emerging materials and related scalable technologies is a highly interdisciplinary field, with the need for researchers, professionals, and academics across the spectrum of engineering and technological disciplines. The main objective of this book series is to attract more attention to these materials and technologies and invite conversation among the international R&D community.

Solar Cells Development and Fabrication
Edited by Shivani Dhall, Kapil Sood and Vinay Gupta

Geopolymers and Composites: Processing Technologies and Applications
Edited by Huirong Le, Kaibao Wang and Longyuan Li

Nanomaterials for Electrochemical Sensing
Edited by Syed Shahabuddin, Rama Gaur and Nandini Mukherjee

Organic-Inorganic Hybrid Nanomaterials: Energy Harvesting, Storage, and Advanced Applications
Edited by Srikanta Moharana, Bibhuti B. Sahu and Suresh Sagadevan

Nanocarbon to Polymer-Nanocarbon Composites: From Synthesis to Breakthrough Performance
Edited by Srikanta Moharana, Bibhuti B. Sahu and Suresh Sagadevan

Materials for Dye Degradation
Edited by Tarun Parangi and Soyebkhan Pathan

Circularity of Cementitious Materials: A Practical Approach
Christian Paglia

Machine Learning for Semiconductor Materials
Edited by Neeraj Gupta, Rashmi Gupta, Rekha Yadav, Sandeep Dhariwal and Rajkumar Sarma

For more information about this series, please visit: www.routledge.com/Emerging-Materials-and-Technologies/book-series/CRCEMT

Machine Learning for Semiconductor Materials

Edited by Neeraj Gupta, Rashmi Gupta, Rekha Yadav, Sandeep Dhariwal and Rajkumar Sarma

CRC Press
Taylor & Francis Group
Boca Raton London New York

CRC Press is an imprint of the
Taylor & Francis Group, an **informa** business

Designed cover image: Shutterstock

First edition published 2026
by CRC Press
2385 NW Executive Center Drive, Suite 320, Boca Raton FL 33431

and by CRC Press
4 Park Square, Milton Park, Abingdon, Oxon, OX14 4RN

CRC Press is an imprint of Taylor & Francis Group, LLC

ISBN: 978-1-032-79688-8 (hbk)
ISBN: 978-1-032-83215-9 (pbk)
ISBN: 978-1-003-50830-4 (ebk)

DOI: 10.1201/9781003508304

Typeset in Times
by Apex CoVantage, LLC

Contents

Preface

The semiconductor industry has grown very rapidly. The electronic system, which is composed of numerous numbers of transistors, is expected to grow similarly. Unfortunately, transistors suffer from process variation, which also results in varying device performance. Low-power consumption and high-speed applications impose a huge requirement on small-sized devices. So metal oxide semiconductor field effect transistors (MOSFETs) have been scaled to meet the market's future requirements. The scaled MOSFET is more prone to process variation as compared to long-channel devices. These scaled devices suffer from short channel effects (SCEs). The device characteristics must be predicted before its fabrication to prevent failure. Currently, the device characteristics with parameter variations are studied by using technology computer-aided design (TCAD). But a long simulation time is required for good accuracy. Recently, a wide range of machine learning algorithms and techniques have been used in every domain of engineering. This approach provides a better trade-off between time and accuracy for meeting market requirements.

One of the most pressing problems in all domains of engineering is the requirement of advanced materials. Conventional materials have approached their limit and cannot be used in future applications. There are various industrial applications where new and advanced materials can play a vital role. This book presents a journey from present to future materials for semiconductors. TCAD is used for device simulation, but it is costly and time consuming to do simulations on TCAD nowadays. Due to TCAD's limitations with device variation and accuracy, machine learning is replacing TCAD. It provides better accuracy and is a powerful technology in material research.

The book *Machine Learning for Semiconductor Materials* studies recent techniques and methods of machine learning to mitigate the use of TCAD, as the use of TCAD tools in semiconductor devices is very tedious. The utilization of machine learning algorithms for semiconductor devices improves device and process accuracy and helps in the analysis of the process variation of the device. This book presents various machine learning algorithms, such as regression, decision tree, support vector machine and *K*-means clustering. It explores the materials and their applications in various domain of science, technology and everyday life to provide further research on the subject. It also highlights semiconductor materials and their uses in multi-gate devices. It is ideal for physicists, computer specialists, engineers, practitioners, researchers, academicians and students who are looking to learn more about machine learning and semiconductor materials and their applications.

Editor Biographies

Neeraj Gupta is an associate professor at Amity University, Haryana, specializing in device modelling, VLSI design, low-power design, analog design, optimization techniques, AI and machine learning and IoT and embedded systems. Dr Gupta received his doctoral degree at Amity University, Haryana, India, in 2019. He has more than 16 years of teaching experience in different subjects. With extensive experience in both academic and research domains, Dr Gupta has contributed significantly to the fields of VLSI circuits and device simulation. His work often bridges cutting-edge technologies with practical applications, making him a key figure in advancing innovations in semiconductor technologies and AI-driven systems. He has published more than 40 international articles in different reputed SCI/SCOPUS indexed journals, presented at conferences, authored two book chapters, and published a book on microprocessors. He also has five published patents. He also received the 2024 Best Researcher and Best Teacher Award from Wegrow Pvt. Ltd.

Rashmi Gupta is an assistant professor at Amity University, Haryana, Gurugram, India. Dr Gupta received her doctoral degree at Amity University, Haryana, India, in 2020. She has more than 13 years of teaching and research experience. Her current research interest is in artificial intelligence, machine learning, software engineering, soft computing, IoT etc. She has published more than 20 international articles in different reputed SCI/SCOPUS journals, two book chapters and a book on software testing. She also has five published patents.

Rekha Yadav is an assistant professor at DCR University of Science and Technology, Murthal, India. She received her doctoral degree from DCR University of Science and Technology, Murthal, Haryana, India, in 2018. She has 15 years of teaching and research experience. Her current research interest is in semiconductor device modelling, VLSI design etc. She has published more than 30 international articles in different reputed SCI/SCOPUS journals and four book chapters.

Sandeep Dhariwal is an associate professor at Alliance University, Bengaluru, India. Dr Dhariwal received his doctoral degree at Banasthali University, Rajasthan, India, in 2015. He has more than 14 years of teaching experience in different subjects. His current research interest is in semiconductor device modelling, low-power CMOS, VLSI design etc. He has published more than 40 international articles in different reputed SCI/SCOPUS indexed journals, presented at conferences, and authored three books. He also has three published patents in multidisciplinary areas of research.

Rajkumar Sarma received his BE in electronics and communications engineering from Vinayaka Mission's University, Salem, India, in 2008. He received his MTech and PhD from Lovely Professional University, Phagwara, Punjab, in 2012 and 2020, respectively. He has more than 11 years of teaching and research experience. He has been working as a postdoctoral researcher at the University of Limerick, Limerick,

Ireland, under the Automatic Design of Digital Circuits (ADDC) project since September 2022. He is also working on a Science Foundation of Ireland (SFI) funded project to generate VHDL/Verilog code for digital circuit design and apply it to the creation of different IP blocks, specifically FIR and FFT filters. His research interests include (but are not limited to) analog and digital VLSI design, low-power architecture design, quantum cellular automata, prototype development using FPGA etc. He has more than 25 research publications in SCI/Scopus indexed and other reputed journals and has presented at national and international conferences. He has published two books on hardware description language and FPGA prototyping. He also has more than 15 published patents in various engineering fields.

Contributors

Manjulata Badi
Department of Electrical and Electronics Engineering
Alliance University
Bengaluru, India

Rajat Butola
Parallel and Scientific Computing Laboratory
Electrical Engineering and Computer Science International Graduate Program
National Yang Ming Chiao Tung University
Hsinchu, Taiwan

Amit Das
Department of Electronics and Communication Engineering
GLA University
Mathura, Uttar Pradesh, India

S. S. Deswal
Department of Electrical and Electronics Engineering
Maharaja Agrasen Institute of Technology
New Delhi, India

Nand Deva
Department of Electronics and Communication Engineering,
Delhi Technological University
New Delhi, India

R. S. Gupta
Department of Electronics and Communication Engineering
Maharaja Agrasen Institute of Technology
New Delhi, India

Charu Jain
Xebia Academy Global
Gurugram, India

Vanshika Jain
Department of CSE
Amity School of Engineering and Technology
Amity University
Haryana, India

Raygun Jose
Xebia Academy Global
Gurugram, India

Deborah T. Joy
Xebia Academy Global,
Gurugram, India

Divia T. Joy
Department of CSE
Amity School of Engineering and Technology
Amity University
Haryana, India

Binod Kumar Kanaujia
Department of Electronics and Communication Engineering
Dr. B. R. Ambedkar National Institute of Technology
Jalandhar, India

Aapurva Kaul
Department of Electronics and Communication Engineering,
Delhi Technological University
New Delhi, India

Prashant Kumar
Department of Electronics and Communication Engineering,
J. C. Bose University of Science and Technology
Faridabad, India

Nivedita Manohar Mathkunti
Department of Electrical and Electronics Engineering,
Alliance University
Bengaluru, India

Arpita Pattnaik
Department of CSE
Amity School of Engineering and Technology
Amity University
Haryana, India

Ila Rai
Department of Electrical and Electronics Engineering
Alliance University
Bengaluru, India

Sonam Rewari
Department of Electronics and Communication Engineering
Delhi Technological University
New Delhi, India

Rajkumar Sarma
Department of Computer Science and Information Systems
University of Limerick
Ireland

Narwal Seema
Department of ECE
Dronacharya College of Engineering
Gurugram, India

Karamjit Kaur Sekhon
Department of ECE
Amity School of Engineering and Technology
Amity University
Haryana, India

Neharika Sharma
Department of ECE
Amity School of Engineering and Technology
Amity University
Haryana, India

C. Shekhar
Department of Physics
Amity School of Applied Sciences
Amity University
Haryana, India

Arun Kumar Singh
Department of CSE
Amity School of Engineering and Technology
Amity University
Haryana, India

Shivani Yadav
Department of Electronics & Communication Engineering
KIET Group of Institutions
Ghaziabad, Uttar Pradesh, India

List of Figures and Tables

Figures

Tables

1 Semiconductor Materials
Current Applications and Limitations of Advanced Semiconductor Devices

C. Shekhar

1.1 INTRODUCTION

Semiconducting materials have played a significant role in the information revolution. The main characteristic of semiconducting materials is the resistivity values falling between those of metals and insulators and that can be tuned with different approaches of material fabrication. The tunable electrical resistivity of these materials has enabled the development of a wide variety of electronic devices that have fundamentally changed the way humans interact among themselves or share, store, and create knowledge and data and changed the world as well. The developments of semiconductor materials started in the 19th century with the discovery of electrical conductivity by Alessandro Volta. Alessandro Volta devised the Voltaic pile, known as the first electrical battery, leading to the development of the study of electrical properties of different types of materials.[1] Following this, in 1833, Michael Faraday studied the corelations between the amount of electricity passing through a solution and the amount of chemical change in the solution and presented the laws of electrolysis, thereby establishing the electron as the carrier of the electrical charge in electrical solutions.[2] Another notable discovery was made in 1873 by Frederick Guthrie, who found the phenomenon of thermionic emission, in which a hot metal rod exerts a force of attraction or repulsion on lightweight objects. Thermionic emission later became the basis of vacuum tubes used to modulate signals for amplification or rectification.[3] The advent of vacuum tubes led to development of electronic devices and the electronic industry. However, due to the inherent disadvantages of vacuum tubes, such as being large, bulky structures, and unreliable output, the demand for more efficient and compact electronic devices kept the search ongoing.[4]

In 1904, John Ambrose Fleming invented the diode, a two-terminal electronic device that allows current to flow in only one direction. The diode was made from a vacuum tube with a heated cathode and cold anode.[5] In 1906, Lee De Forest invented the triode, a three-terminal electronic device that could be used to amplify electrical signals. The triode was made from a vacuum tube with a heated cathode, cold anode, and control grid. Vacuum tubes were used in a variety of applications, including

DOI: 10.1201/9781003508304-1

radio, television, and radar. However, they were still large, bulky, and unreliable. This led to a search for a more efficient and compact electronic device.[6]

The continued effort and search to replace vacuum tubes with more efficient, compact, and reliable electronic devices led J. Bardeen, W. Brattain, and W. Shockley in 1947 to develop transistors. The transistor is a semiconductor device capable of amplifying and switching electrical signals made by arranging extrinsic semiconductors in PNP or NPN format. Due to its improved properties, the transistor replaced vacuum tubes in different applications and fuelled the proliferation of a variety of electronic devices.[7,8] This further led to the invention of the integrated chip (IC) by J. Kilby and R. Noyce in 1950, which is a compact electronic device housing thousands of transistors and other electronic structures.[9–11] With the development of ICs, it became possible to design and fabricate highly complex electronic structures at a reduced cost, leading to the development of power-efficient, compact electronic devices with greater computing power, thus giving rise to the modern electronic revolution. Such electronic devices are used in everything from computers to watches, missiles, aircraft, and smartphones.

1.2 TRADITIONAL SEMICONDUCTORS

1.2.1 Silicon (Si)

Silicon (Si) is among the most abundant materials in the earth's crust, making up almost 27% of it. Additionally, it has favourable electronic properties, which can be tuned by different approaches. Its semiconducting nature is most useful for the electronic industry; therefore, it has been most widely used for the manufacturing of electronic components for the electronic industry. The properties of silicon have paved the way for it being used to fabricate many electronic components for use in a multiplicity of applications; therefore, silicon has dominated the semiconductor industry for many years. Among the many components fabricated on a large scale from silicon are transistors, diodes, ICs, sensors, and solar cells, to mention a few. This section highlights the significance of silicon as the material of choice for the semiconductor industry along with some of its prominent limitations for advanced applications.[12]

After oxygen, silicon is the most abundant material present on the earth. It is generally extracted from silica sand. Its abundance and availability make it a suitable candidate for large-scale use. High-quality silicon crystals are grown from the solution obtained from silicon dioxide, which is needed for the fabrication process in the semiconductor industry. With the low bandgap of 1.12 eV, silicon allows the effective control of charge carriers at room temperature, and with suitable doping, the carrier concentrations can be controlled by adding aliovalent cations to the Si lattice. This ability to effectively control carrier concentrations, and therefore the conductivity of Si, makes it a suitable candidate for applications such as transistors, switching devices, and amplifiers. When silicon is exposed to oxygen, it forms an insulating layer of SiO_2 readily, which is critical for the fabrication of metal oxide semiconductor field effect transistors (MOSFETs). MOSFETs have been central to many electronic components in digital and analog circuits. Additionally, SiO_2, formed on exposure of silicon to oxygen, provides much-needed electrical insulation between different components when desired, enhancing the reliability of the electronic components for prolonged usage.[13]

Since its first development a long time ago, Si-based industries have created large-scale synthesis and fabrication capabilities – particularly, the Si-based ICs and complementary metal oxide semiconductor technology, enabling the creation of millions of transistors on a single chip for high-performance computing. The mature infrastructure capabilities, skilled workforce, and cost considerations are some of the reasons for Si being the dominant material over the years in the semiconductor industry.

1.2.2 Limitations of Silicon for Advanced Semiconductor Devices

Though silicon possesses many admirable properties, it has significant limitations, especially in the areas of high-frequency, high-speed devices. Let us look at some of the limitations briefly here: (1) Compared to gallium arsenide (GaAs), the mobility of the charge carrier in silicon is low. For applications requiring high-frequency operations, such as radio-frequency (RF) and microwave devices, materials with a large charge carrier mobility are more efficient. Therefore, materials such as GaAs are preferred over Si for high-speed operation at lower power consumption. (2) Silicon is not the preferred material in areas of power electronics due to its lower thermal conductivity compared to other materials, such as silicon carbide (SiC) or diamond. In high-power applications, the ability to manage the heat generated becomes crucial for its efficient dissipation. Materials such as SiC and diamond are among the preferred materials in power electronics due to better thermal conductivity and higher breakdown voltage. (3) With the advancement of science, particularly in the area of nanotechnology, Si faces challenges at the nanoscale, pertaining to quantum tunnelling and short-channel effects. These effects lead to power leakage and reduced performance for ultra-small transistors.[14] Due to these limitations, the scientific community is looking for other 2D materials, such as graphene, transition metal dichalcogenides, and III-V semiconducting compounds, for next-generation devices.

1.2.3 Future Prospects and Alternatives

There is now a class of different semiconducting materials suitable for replacing silicon, such as SiC and gallium nitride (GaN). Though Si remains the preferred material of choice in many areas, there is a gradual shift in advanced areas, such as high-power and high-frequency applications. Due to the existing infrastructure, its cost-effectiveness, and wide range of applicability, Si is still greatly favoured.

1.3 TRADITIONAL SEMICONDUCTORS: GERMANIUM (GE)

Another material earlier used for the fabrication of electronic devices is germanium (Ge). It was first used for the preparation of transistors in the early 1940s and 1950s. Transistors made of Ge were widely used before being replaced by Si. Although Ge has been replaced by Si in a majority of areas, it is still the preferred choice in high-frequency and optoelectronic applications due to its superior charge carrier mobility and direct bandgap, respectively.

With a lower bandgap of 0.66eV, Ge can conduct at close to room temperature as compared to Si, which can only conduct at higher temperatures due to a wider bandgap. This makes it a more suitable candidate for many electronic devices such as diodes and transistors. At the same time, this also means that electronic devices made of Ge are more sensitive to temperature variations, leading to higher leakage currents compared to Si.[13,15] In comparison to silicon, the charge carrier mobility of germanium is significantly higher. The electron and hole mobilities in germanium and silicon are 3,900 $cm^2/V{\cdot}s$ and 1,900 $cm^2/V{\cdot}s$ and ~1500 $cm^2/V{\cdot}s$ and ~450 $cm^2/V{\cdot}s$, respectively. The higher charge carrier mobility suggests the suitability of germanium in high-frequency applications, in which rapid switching and signal processing are key requirements. Though the charge carrier mobility is significantly higher (149 W/mK) for germanium, its thermal conductivity is lower (60 W/mK). Thus, germanium faces issues in managing the heat generated in power electronics and high-temperature applications.

Before the widespread use of germanium, vacuum tubes were used for the amplification of signals. The use of Ge revolutionized the process and led to the fabrication of the first bipolar junction transistors in the 1940s. This development enabled the creation of compact and energy-efficient amplifiers, replacing the vacuum tubes. Till the 1960s, Ge was dominant; it was used to make a wide variety of electronic devices from radio to the early computers. But the sensitivity of Ge for higher temperatures and its relatively higher cost compared to Si eventually led to silicon replacing it. The ability of silicon to form SiO_2 with oxygen led to the development of MOSFETs, which laid the foundations of the modern electronic industry.[16]

1.3.1 Current Applications and Limitations

Although Ge was replaced by Si in many applications, its use in many applications is still mainstream, as described here:

- **High-Speed Transistors:** The higher carrier mobility of Ge is suitable for high-frequency and high-speed transistors, considering the switching frequency and power consumption requirements.
- **Optoelectronics:** The lower bandgap of Ge is suitable for the detection and measurement of low-energy radiations, such as infrared (IR) and fibre optic communication systems.
- **Solar Cells:** Ge can enhance the energy efficiency of the substrate in multi-junction solar cells and also has space applications.
- Alloys and compounds: Ge is very often alloyed with Si; for example, SiGe alloy is used in RF and microwave applications.

Regardless of these advantages, Ge has limitations, such as being expensive to process, lower thermal conductivity, and inability to form an oxide layer like SiO_2, all of which limit its applications and render it less versatile in MOSFET applications.

1.3.2 Future Prospects

With the growing demand for compact, faster, and more energy-efficient devices, there is renewed interest in Ge-based electronic devices to overcome the scaling limitations of Si. Moreover, there are efforts to investigate Ge's use in quantum computing and photonic devices, where its unique electrical and optical properties may offer additional advantages. There are therefore possibilities for the use of Ge in new and specialized ways in electronic devices.

1.4 GALLIUM ARSENIDE (GaAs)

GaAs is a direct bandgap, III-V semiconductor. It has been used in critical applications, such as microwave and millimetre-wave devices, photovoltaic cells, and light-emitting diodes (LEDs). Being a direct bandgap semiconductor and a key material in the electronics industry for a long time, it offers unique advantages over traditional materials, such as Si, particularly in high-frequency, high-speed, and optoelectronic applications.

The bandgap of GaAs is 1.42 eV at room temperature, whereas Si is an indirect bandgap material. This makes GaAs more energy efficient for electron transitions from the conduction band to the valence band and vice versa, leading to effective photon emission. Thus, GaAs is widely used in optoelectronic applications such as laser diodes and LEDs.

The charge carrier mobility of GaAs is ~8,500 $cm^2/V{\cdot}s$, which is significantly higher compared to Si (1,500 $cm^2/V{\cdot}s$). The higher carrier mobility translates to a quick response to the applied external field, making GaAs more suitable for high-frequency, high-speed applications, such as microwave transistors and field effect transistors (FETs). GaAs-based devices can be effectively used in the GHz frequency range, which is suitable for RF and microwave communication systems. The higher stability of GaAs-based devices against radiation damage also places these devices ahead of Si-based devices for space and satellite applications, where the performance of the latter can degrade significantly. GaAs-based devices show higher performance stability for longer periods in harsh conditions.[17]

1.4.1 Applications of GaAs in Semiconductor Devices

Due to the low direct bandgap, high carrier mobility, and high saturation velocity, GaAs-based devices have been extensively used in high-frequency and optoelectronic applications. GaAs-based devices – for example, microwave and millimetre-wave transistors such as heterojunction bipolar transistors and high-electron-mobility transistors – have been widely used for many applications, like telecommunications, radar, wireless communication systems, optoelectronics, solar cells, LEDs, and laser diodes. Photovoltaic systems such as GaAs-based solar cells have been used in satellite and space applications due to their high efficiency, durability, and high performance under extreme radiation exposure for long periods of time. Also, GaAs is the preferred material for the fabrication of monolithic microwave integrated Circuits (MMICs), integral for radar, satellite, and cellular-based communication systems.

Additionally, due to its direct bandgap, GaAs is ideal for optoelectronic applications such as efficient LEDs and laser devices.[18]

1.4.2 Limitations of Gallium Arsenide

Although there are many advantages of GaAs, its drawbacks have limited its application in many areas. GaAs is expensive to synthesize compared to Si primarily due to the scarcity of Ga. The wafer fabrication of GaAs is complex, increasing the overall cost and limiting its use to specialized high-performance applications. Its mechanically fragile nature makes it difficult to handle during the fabrication process, thus making it challenging to scale the production of GaAs-based systems. Its lower thermal conductivity limits its application where heat dissipation is critical for device performance. Due to these issues, SiC and GaN are preferred instead in many applications.

1.4.3 Future Prospects

Although GaAs is critical for many applications, materials like SiC and GaN are also used for power electronics and high-performance transistors. GaAs continues to be critical for high-frequency and optoelectronic applications where efficient light emission and stability against solar radiation are critical requirements.

1.5 ADVANCED SEMICONDUCTOR MATERIALS

With the advancement of science and technology and electronic devices, in particular, the demand for compact, energy-efficient devices have increased in recent years. The properties of traditional semiconducting materials limit their applications in many areas, such as power electronics, telecommunications, and optoelectronics. These limitations have led to the development of newer materials with superior physical and chemical properties for such applications. Among other materials, silicon carbide (SiC), gallium nitride (GaN), and indium phosphide (InP) are useful and promising candidates for engineering and technological applications.

1.6 SILICON CARBIDE (SIC)

1.6.1 Properties and Advantages

Silicon carbide – the wide-band semiconductor with a bandgap of 3.26 eV, high breakdown voltage, high thermal conductivity, mechanically hard, and thermal and structural stability at high temperatures and frequencies – is a suitable candidate for devices that operate at high frequencies and high voltages in the area of power electronics. Its high thermal conductivity (~4.9 W/cm·K) ensures efficient thermal management in high-temperature applications without the need for detailed cooling supports for the device, making such devices rugged enough to function under harsh conditions. The devices made of SiC, such as MOSFETs and Schottky diodes, show much lower ON-state resistance and switching losses as compared to traditional semiconductor-based devices, leading to drastically improved efficiencies. These characteristics are critical for many energy-sensitive applications, such as electric vehicles (EVs), renewable energy systems, and industrial motor drives.[19]

1.6.2 Challenges and Current Developments

Though there are significant advantages with the properties of SiC for applications in critical areas, there are limitations too: (1) The cost and complexity of SiC-based devices is way higher than that for the Si-based devices. (2) SiC is a superhard and brittle material and requires special care to polish and cut the crystal into fibres for the fabrication processes. Thus, the wafer manufacturing is a costly and complex process. The technology for the fabrication of SiC wafers is not yet as mature as for Si-based devices.

To address these challenges and harness the unique properties of SiC, there have been sustained efforts to develop technologies to circumvent its limitations. Some of the efforts centred on developing physical vapour deposition (PVD), chemical vapour deposition (CVD), and single crystal growth processes to improve the availability of SiC crystals and wafers. These improvements have led to the enhanced usage of the material for commercial applications. Additionally, efforts for a new electronic architecture development and fabrication process are in focus. This has resulted in trench MOSFET structures, created to minimize ON-state resistance and improve the performance of SiC-based devices.

1.6.3 Applications

The primary application of SiC is in power electronics, where it is used in high-voltage, high-temperature, and high-efficiency devices. SiC MOSFETs and Schottky diodes are increasingly being used in power converters, inverters, and motor drives due to their ability to operate at higher switching frequencies and temperatures compared to silicon-based devices. This enables more compact, efficient, and reliable power systems, especially in EVs, industrial equipment, and renewable energy installations.

In the automotive industry, SiC is being used in the powertrains of electric and hybrid vehicles to improve efficiency and reduce the size and weight of power electronics systems. The high thermal conductivity of SiC allows for more effective heat dissipation, reducing the need for large cooling systems and enabling more compact designs.

SiC is also finding applications in aerospace and defence, where its ability to withstand extreme temperatures and radiation makes it suitable for use in harsh environments. SiC devices are also used in satellite communication systems, radar, and high-temperature sensors.

1.7 GALLIUM NITRIDE (GAN)

1.7.1 Properties and Advantages

With a bandgap of 3.4 eV, gallium nitride is a preferred choice for many applications. The bandgap of GaN is even greater than that of SiC; it has high carrier mobility and saturation velocity, enabling its use in high-frequency applications, and GaN-based devices work at speeds much greater than that for Si-based devices. Due to these properties, GaN can be used for a multiplicity of applications involving high frequency and high power, such as RF, power transistor, and high-speed communication systems.

Gallium nitride has a high breakdown voltage and thermal stability. It is stable at high voltages and elevated temperatures. Due to its high thermal conductivity, it can efficiently operate at higher temperatures without the need for cooling systems. In addition, GaN crystals have high charge carrier mobility. Also, due to the compatible lattice parameters, GaN can be grown on silicon. It therefore can utilize the existing fabrication and manufacturing techniques for silicon, enabling the integration of technology tailored for Si to be used for GaN.[20]

1.7.2 Challenges and Current Developments

There are many challenges to the development of GaN as the mainstream electronic material as compared to Si primarily due to the complexity of the manufacturing process involved. The cost of the manufacturing process is larger as compared to that of Si, thus limiting GaN's applicability. There are reliability issues in high-power applications, and GaN-based devices are prone to degradation in high-stress conditions.

Among the current developments in GaN-based applications, they have shown improved performance in the area of high switching speeds with low losses. The emphasis is on the development of compact electronic devices with better design and packaging techniques and improved efficiencies. Recent developments have focused on their applications in electric mobility, improving the power efficiency of charging stations, and enhancing the range of electrical vehicles.

1.7.3 Applications

GaN is widely used in high-frequency and high-power applications, such as RF amplifiers, power transistors, and microwave devices. Its high electron mobility and saturation velocity make it ideal for use in high-frequency communication systems, including 5G networks and satellite communication.[21]

In the power electronics sector, GaN devices are being used in power converters, inverters, and motor drives, where their high efficiency and fast switching speeds lead to significant energy savings. GaN is also being used in power supplies for data centres and telecommunications equipment, where its ability to handle high voltages and operate at high frequencies helps to reduce power consumption and heat generation.[22] GaN is also finding applications in optoelectronics, particularly in LEDs and laser diodes. GaN-based LEDs are widely used in solid-state lighting, display technology, and automotive lighting due to their high efficiency and brightness.

1.8 ADVANCED SEMICONDUCTOR MATERIALS: 2D MATERIALS – GRAPHENE AND TRANSITION METAL DICHALCOGENIDES

There have been tremendous advances in information technology based on the availability of better products that are used for the storage, transfer, and processing of information in enormous quantities. This has been possible primarily due to the availability of semiconductors capable of operating at these scales. The present density of transistors, their energy efficiency, operating temperatures, thermal management capacities, and operational frequencies are at previously unimaginable levels. Due to the ever-increasing demand for better products, there is a requirement for materials with improved properties that can extend the limits of present-day electronic devices. In the recent past, there has been significant development in the areas of 2D materials such as graphene and transition metal dichalcogenides (TMDs), which can be promising candidates due to their excellent physical and chemical properties that are crucial for their use in electronic devices.

The much-famed 2D materials consist of a two-dimensional extended layer of atoms with the thickness of a few atomic layers. This leads to a dramatic alteration of the properties of this material due to the quantum confinement effects in the 2D materials. Among 2D materials, graphene is the most important material reported in 2004, which is a single-layer thick carbon sheet arranged hexagonally. In addition, TMDs such as molybdenum disulfide (MoS_2) and tungsten disulfide (WS_2) have attracted great attention because of their superior semiconducting properties.

1.9 GRAPHENE: THE PIONEER OF 2D MATERIALS

1.9.1 Discovery and Structure

Graphene was prepared by A. Gein and K. Novoselov in 2004, and they won the 2010 Nobel Prize in Physics for the discovery of this remarkable material. It consists of a hexagonal arrangement of carbon atoms in a single layer, creating an electronic band structure that results in unique physical and chemical properties. Graphene has zero bandgap – that is, it's a semimetal. The topmost of the valance band and lowest of the conduction band touches at the Dirac point, resulting in the massless Dirac fermions, thus imparting excellent physical and chemical properties to graphene.[23]

1.9.2 Properties

The single-layer carbon, arranged in a hexagonal pattern forming the zero bandgap material, has excellent physical properties, which are extremely useful in electronic device applications. Graphene exhibits enormous electron mobility of 2×10^5 cm^2/Vs under ideal conditions.[24] Such a value of electron mobility is remarkably high compared to traditional conductors such as silver and copper. Graphene is extremely strong, with 130 GPa tensile strength, almost 200 times that of steel.[24] Its thermal conductivity is of the order of 5,000 W/mK, making it the ideal candidate for thermal management in electronic devices.[25] It only absorbs a very small portion of the electromagnetic radiation incident on it. Graphene is a highly transparent material and suitable for optoelectronic applications.[26] Although is very strong compared to steel, graphene is flexible at the same time and can be stretched by almost 20%, making it the best candidate for flexible electronics.

1.9.3 Applications

The high mobility of graphene has made it a most suitable candidate for replacing traditional electronic materials to make FETs, and devices made of graphene can be used at high signal frequencies, leading to devices with much faster speeds. The large surface area and high electrical conductivity of graphene make it suitable for use in high-storage batteries and supercapacitors, significantly increasing the energy density of these devices.[27] It is also useful for surface-sensitive applications such as sensors for gas detection, chemicals, and other molecules.[28] The flexible nature of graphene makes it a much desired candidate for flexible electronics, wearables, and foldable displays, to mention a few applications.[29] In addition to this, the transparent nature of graphene has many applications in optoelectronic devices, such as LEDs transparent electrodes, and photodiodes.[30]

1.9.4 Challenges and Recent Advances

Although graphene has excellent properties for application in different areas, there are inherent limitations. The main limitation is the lack of the bandgap, in that it that shortens its applicability in electronic devices. It is challenging for the scientific community to find ways to synthesize graphene in a reproducible manner and create a bandgap in graphene using various approaches. Additionally, there are efforts to intercalate graphene with other promising materials such as TMDs to create composites and heterostructures that can address some of the pressing issues and explore the possibilities for a new class of suitable materials.[31]

1.10 TRANSITION METAL DICHALCOGENIDES

1.10.1 Structure and Composition

Two-dimensional TMDs have the formula MX_2 (M = Mo, W; X = S, Se, Te). TMDs, unlike graphene, are semiconducting and therefore are suitable for electronic applications. Unlike other materials, TMDs can be easily exfoliated into monolayers. Their properties vary from material to material; for example, MoS_2 is a low-bandgap (1.8 eV) material, making it suitable for different applications such as transistors and optoelectronic devices.[32]

1.10.2 Properties

The 2D TMDs in their monolayer form are direct bandgap semiconductors, which are indirect bandgap semiconductors in the bulk form. Therefore, they are an excellent candidate for optoelectronic applications.[33–35] The monolayer form of TMDs is flexible and can be used in flexible electronic devices. The reasonably high electronic mobilities of the charge carriers in TMDs is good for their use in thin-film transistors.[36]

1.10.3 Applications

TMDs are low-bandgap materials, and therefore, they can absorb electromagnetic radiations of a lower energy. This makes them suitable candidates for photoluminescence, photodiodes, and other optoelectronic applications.[35] Transition metal oxides have been in focus for their applications in electronics, such as FETs, high-switching devices, and power-efficient devices.[37]

Additionally, the 2D TMDs are mechanically flexible and suitable for wearable electronics as a result of their suitable electronic properties.[37] The thickness-dependent properties of TMDs have been exploited to design nonvolatile memories to enhance the performance of electronic devices.[38] TMDs have also found application in the emerging areas of catalysts in hydrogen evolution and storage solutions for a sustainable future.[39]

1.10.4 Challenges and Recent Advances

TMDs have proven themselves to be promising materials for enhancing the functionality of electronic devices. But there are challenges, on which much progress is needed to encash the full potential of these promising materials. The main limiting factor is the challenge to scale the laboratory findings to an industrial scale to ramp up the production of TMDs-based electronic devices. These materials show excellent properties in small lab settings, but in larger areas, due to the variation in the thickness of these materials, a variation in the resultant properties has been observed. Another area of concern is the metal electrode and TMD interface/contact resistance in TMD-based devices. The resistance at the interface can adversely affect the performance of electronic devices. There have been efforts to grow large-area TMDs with uniform properties and minimize the contact resistance between TMDs and metal electrodes. Another approach is the development of heterostructures with graphene and other TMDs for advancing the use of heterostructures in multifunctional devices.[40]

1.11 LIMITATIONS OF TRADITIONAL MATERIALS FOR MODERN DEVICES: SCALING CHALLENGES IN THE CONTEXT OF MOORE'S LAW

In the 1960s, Moore presented an empirical law by which he predicted the advancement in the electronic industry and the revolutionizing of the ways of human interaction and knowledge and data creation, handling, and transmission. According to Moore's law, the number of transistors on a chip would double every two years. The relentless development of electronic devices has been keeping pace with the prediction of Moore's law for over six decades now. The exponential rise in the ability to integrate a considerable number of electronic components on a chip has led to electronic devices with higher computational power and lower costs and energy requirements at the same time. But as we are reaching the limits of integrating electronic components using traditional semiconducting materials, the need for more space and energy-efficient materials is growing. In the next section, we will delve into the limitations of traditional semiconducting materials vis-à-vis Moore's law and examine its effects on semiconductor technology.

1.12 LIMITATIONS OF SILICON IN MODERN DEVICES

1.12.1 Physical Scaling Limits of Silicon

As the size of electronic components decreases to sub-10 nm in electronic devices, quantum mechanical effects start taking prominence in the densely packed ICs. The quantum tunnelling effect is more pronounced when the thickness of the insulating layer is of the order of tens of nanometres in a transistor. As a result, the electron may tunnel across the insulating SiO_2 layer in the case of Si wafers, creating uncertain pathways of electron motion in the transistor, leading to increase in the leakage current. Quantum tunnelling compromises the switching ability of the transistor to turn off, thus increasing power requirements and lowering the efficiency of electronic devices. With the decrease in thickness of gate length, the ability of the electron to flow from the gate to the drain using gate voltage decreases, thus increasing the leakage and diminishing the performance of electronic devices.

1.12.2 Power Density and Thermal Dissipation

With the increase in the density of components on a chip, the power density also increases, leading to challenges in thermal management. Due to the thermal conductivity of traditional semiconductors, there are issues in the management of heat generated during device operation. Additionally, the quantum effects that lead to the tunnelling of the charge carrier across the insulating barrier, especially in high-frequency operations, significantly contribute to heat generation. The common and traditional methods employed to cool electronic devices have their limitations, which can cause heat management issues in the devices and create performance instabilities.

1.12.3 Electrical Performance and Interconnect Issues

As the size of the component decreases, the performance of the metal wires, known as the interconnects, that connect different transistors becomes significant. With the decrease in the thickness of the interconnects, the resistivity of interconnects made of copper or aluminium increases, creating a critical bottleneck for charge transfer. Also, with the increase in the density of interconnects, the capacitance between the tightly packed interconnects increases. The combined effect of the increased resistivity and capacitance leads to resistive-capacitive (RC) delay in the circuit and many a time negates the increase in signal speed achieved for smaller components.

1.13 THE FUTURE OF SEMICONDUCTOR TECHNOLOGY

There are many challenges for the semiconductor industry to keep up with the pace of development as envisaged by Moore's law. There are scaling challenges for the future of the electronic industry, and there are opportunities to meet those challenges with innovative approaches by employing new materials, device architectures, and manufacturing techniques. The use of new materials and technologies will change the industry and enable the application of these materials in the areas of artificial intelligence, quantum computing, and autonomous systems. The heterogeneous integration of different systems into the same chip will help overcome the scaling limitations and allow the utilization of the properties of dissimilar materials to improve the performance of electronic devices. Quantum computing will open new avenues for the application of advanced technologies with greater computational power to perform intricate calculations in the frontier areas of science and technology.

1.14 CONCLUSION

Semiconductor materials have played a vital role in the development of modern technology. They have enabled the creation of devices that have transformed our world, from computers and smartphones to solar cells and LEDs. The future of semiconductor materials is bright, with new and exciting applications being developed all the time.

Silicon's role as a traditional semiconductor material is deeply ingrained in the electronics industry. Its favourable properties, such as availability, moderate bandgap, and compatibility with MOSFET fabrication, have enabled its long-standing dominance. However, as technology progresses towards higher-performance applications, materials with better electronic and thermal properties are being explored for specialized uses. Despite this, silicon is expected to continue being a cornerstone of semiconductor technology for many years to come, especially in general-purpose computing and electronics.

Germanium, as one of the original semiconductor materials, played a crucial role in the early development of the electronics industry. While silicon has since overtaken germanium for most applications, germanium's superior electron and hole mobility, along with its infrared sensitivity, ensure that it remains relevant in high-speed, optoelectronic, and specialized applications. As innovative technologies emerge, germanium continues to be a valuable material for advanced semiconductor devices.

Gallium arsenide is a traditional semiconductor material that has proven invaluable in various advanced electronic and optoelectronic applications. Its direct bandgap, high electron mobility, and radiation resistance have made it a preferred material for high-frequency devices, MMICs, and space-based solar cells. While it faces competition from other semiconductor materials in certain areas, GaAs continues to hold a vital position in the semiconductor industry, particularly where high performance and specialized functionality are required.

Silicon carbide, gallium nitride, and indium phosphide represent the next generation of semiconductor materials, each offering unique advantages for high-power, high-frequency, and optoelectronic applications. SiC is ideal for power electronics, where its high efficiency and thermal stability enable more compact and reliable devices. GaN excels in high-frequency and high-power applications, making it a key material for RF amplifiers, power transistors, and communication systems. InP is the material of choice for high-speed and optical communication systems, where its high electron mobility and direct bandgap enable efficient data transmission.

While these materials present challenges in terms of cost and manufacturing, ongoing research and development are driving improvements in their performance and scalability. As the demand for more efficient, faster, and more reliable semiconductor devices continues to grow, SiC, GaN, and InP are poised to play a critical role in shaping the future of electronics and communication technologies.

Graphene and TMDs represent the forefront of advanced semiconductor materials. Graphene's unparalleled conductivity, strength, and flexibility make it a strong candidate for various electronic, optical, and mechanical applications. However, its lack of a bandgap limits its use in digital electronics. On the other hand, TMDs offer the semiconducting properties required for switching devices, along with flexibility and strong photoluminescence, positioning them as prime candidates for next-generation electronics and optoelectronics.

As research continues, the combination of graphene, TMDs, and other 2D materials in heterostructures is likely to lead to breakthroughs in device performance and functionality. While challenges such as scalability and contact engineering remain, the future of these materials in the semiconductor industry is promising.

The limitations of traditional materials, particularly silicon, pose significant challenges for the continued scaling of semiconductor devices. As transistor sizes approach the atomic scale, quantum mechanical effects, power dissipation, and economic constraints are making it increasingly difficult to maintain the pace predicted by Moore's law. However, the exploration of alternative materials such as silicon carbide, gallium nitride, and 2D materials offers promising solutions to these challenges.

In the coming decades, the semiconductor industry will likely transition to a post Moore's law era, where innovations in materials, device architectures, and quantum computing will drive further advancements in technology. While the road ahead is uncertain, the potential for breakthroughs in semiconductor technology remains immense, ensuring that the next generation of devices will continue to push the boundaries of what is possible.

REFERENCES

1. Volta, A. (1800). "On the electricity excited by the mere contact of conducting substances of different kinds." *Philosophical Transactions of the Royal Society of London*, 90, 403–431. https://doi.org/10.1098/rstl.1800.0018
2. Faraday, M. (1834). "Experimental researches in electricity. Seventh series." *Philosophical Transactions of the Royal Society of London*, 124, 77–122. https://doi.org/10.1098/rstl.1834.0004
3. Guthrie, F. (1873). "On a relation between heat and static electricity." *Philosophical Magazine*, 45(Series 4), 257–266.
4. Payen, F. (2001). "An analysis of the advantages and limitations of vacuum tubes in terms of performance and reliability." *REE, Revue de L'Electricite et de L'Electronique*, (4), 28–34.
5. Fleming, J. A. (1905). "On the conversion of electric oscillations into continuous currents by means of a vacuum valve." *Proceedings of the Royal Society of London. Series A, Containing Papers of a Mathematical and Physical Character*, 74(505), 476–487.
6. De Forest, L. (1907). "Device for amplifying feeble electrical currents." U.S. Patent No. US841387A. Filed October 25, 1906.
7 Waldrop, M. M. (2016). "The chips are down for Moore's law." *Nature*, 530(7589), 144–147.
8. Moore, G. E. (1965). "Cramming more components onto integrated circuits." *Electronics*, 38(8).
9. Iwai, H. (2009). "Roadmap for 22nm and beyond." *Microelectronics Engineering*, 86(7–9), 1520–1528.

10. Colinge, J.-P. (2007). *FinFETs and Other Multi-Gate Transistors*. Springer Science & Business Media.
11. Gabor, N. M., Song , J. C., Ma, Q., Nair, N. L., Taychatanapat, T., Watanabe, K., Taniguchi, T., Levitov, L. S., & Jarillo-Herrero, P. (2011). "Hot carrier-assisted intrinsic photo response in graphene." *Science*, 334(6056), 648–652.
12. Streetman, B. G., & Banerjee, S. (2005). *Solid State Electronic Devices* (6th ed.). Pearson Prentice Hall.
13. Sze, S. M., & Ng, K. K. (2007). *Physics of Semiconductor Devices* (3rd ed.). Wiley-Interscience.
14. Mönch, W. (2004). *Semiconductor Surfaces and Interfaces* (3rd ed.). Springer.
15. Muller, R. S., & Kamins, T. I. (2003). *Device Electronics for Integrated Circuits* (3rd ed.). Wiley.
16. Hull, R. (Ed.). (1999). *Properties of Crystalline Germanium*. INSPEC, Institution of Electrical Engineers.
17. Pal, A., & Basu, S. (2017). *Gallium Arsenide Electronics*. Springer.
18. Pankove, J. I. (1971). *Optical Processes in Semiconductors*. Prentice-Hall.
19. Neudeck, P. G., Okojie, R. S., & Chen, L.-Y. (2002). "High-temperature electronics – a role for wide bandgap semiconductors?" *Proceedings of the IEEE*, 90(6), 1065–1076.
20. Baliga, B. J. (2013). "Gallium nitride (GaN) technologies for power electronic applications: Status and future prospects." *Journal of Electronic Materials*, 42(6), 896–903. https://doi.org/10.1007/s11664-012-2471-0
21. Mishra, U. K., Parikh, P., & Yi-Feng, W. (2002). "AlGaN/GaN HEMTs – an overview of device operation and applications." *Proceedings of the IEEE*, 90(6), 1022–1031. https://doi.org/10.1109/JPROC.2002.1021567
22. Musumeci, S., & Barba, V. (2023). "Gallium nitride power devices in power electronics applications: State of art and perspectives." *Energies*, 16(9), 3894.
23. Novoselov, K. S., Fal'ko, V. I., Colombo, L., Gellert, P. R., Schwab, M. G., & Kim, K. (2012). "A roadmap for graphene." *Nature*, 490(7419), 192–200.
24. Lee, C., Wei, X., Kysar, J. W., & Hone, J. (2008). "Measurement of the elastic properties and intrinsic strength of monolayer graphene." *Science*, 321(5887), 385–388.
25. Balandin, A. A., Ghosh, S., Bao, W., Calizo, I. , Teweldebrhan, D., Miao, F., Lau, C. N. (2008). "Superior thermal conductivity of single-layer graphene." *Nano Letters*, 8(3), 902–907.
26. Nair, R. R., Blake, P., Grigorenko, A. N., Novoselov, K. S., Booth, T. J., Stauber, T., Peres, N. M. R., & Geim, A. K. (2008). "Fine structure constant defines visual transparency of graphene." *Science*, 320(5881), 1308.
27. Stoller, M. D., Park, S., Zhu, Y., An, J., & Ruoff, R. S. (2008). "Graphene-based ultracapacitors." *Nano Letters*, 8(10), 3498–3502.
28. Schedin, F., Geim, A. K., Morozov, S. V., Hill, E. W., Blake, P., Katsnelson, M. I., & Novoselov, K. S. (2007). "Detection of individual gas molecules adsorbed on graphene." *Nature Materials*, 6(9), 652–655.
29. Kim, K. S., Zhao, Y., Jang, H., Lee, S. Y., Kim, J. M., Kim, K. S., Ahn, J-H., Kim, P., Choi, J-Y., & Hong, B. H. (2009). "Large-scale pattern growth of graphene films for stretchable transparent electrodes." *Nature*, 457(7230), 706–710.
30. Bonaccorso, F., Sun, Z., Hasan, T., & Ferrari, A. C. (2010). "Graphene photonics and optoelectronics." *Nature Photonics*, 4(9), 611–622.
31. Geim, A. K., & Grigorieva, I. V. (2013). "Van der Waals heterostructures." *Nature*, 499(7459), 419–425.
32. Splendiani, A., Sun, L., Zhang, Y., Li, T., Kim, J., Chim, C-Y., Galli, G., & Wang, F. (2010). "Emerging photoluminescence in monolayer MoS_2." *Nano Letters*, 10(4), 1271–1275.
33. Wang, Q. H., Kalantar-Zadeh, K., Kis, A., Coleman, J. N., & Strano, M. S. (2012). "Electronics and optoelectronics of two-dimensional transition metal dichalcogenides." *Nature Nanotechnology*, 7(11), 699–712.

34. Mak, K. F., Lee, C., Hone, J., Shan, J., & Heinz, T. F. (2010). "Atomically thin MoS_2: A new direct-gap semiconductor." *Physical Review Letters*, 105(13), 136805.
35. Lopez-Sanchez, O., Lembke, D., Kayci, M., Radenovic, A., & Kis, A. (2013). "Ultrasensitive photodetectors based on monolayer MoS_2." *Nature Nanotechnology*, 8(7), 497–501.
36. Radisavljevic, B., Radenovic, A., Brivio, J., Giacometti, V., & Kis, A. (2011). "Single-layer MoS_2 transistors." *Nature Nanotechnology*, 6(3), 147–150.
37. Pu, J., Yomogida, Y., Liu, K-K., Li, L-J., Iwasa, Y., & Takenobu, T. (2012). "Highly flexible MoS_2 thin-film transistors with ion gel dielectrics." *Nano Letters*, 12(8), 4013–4017.
38. Yu, Z., Pan, Y., Shen, Y., Wang, Z., Ong, Z-Y., Xu, T., Xin, R., Pan, L., Wang, B., Sun, L., Wang, J., Zhang, G., Zhang, Y. W., Shi, Y., & Wang, X. (2013). "Towards intrinsic charge transport in monolayer molybdenum disulfide by defect and interface engineering." *Nature Communications*, 5, 5290.
39. Allain, A., Kang, J., Banerjee, K., & Kis, A. (2015). "Electrical contacts to two-dimensional semiconductors." *Nature Materials*, 14(12), 1195–1205.
40. Novoselov, K. S., Mishchenko, A., Carvalho, A., & Castro Neto, A. H. (2016). "2D materials and van der Waals heterostructures." *Science*, 353(6298), aac9439.

2 Machine Learning
Introduction and Features

Deborah T. Joy, Raygun Jose, Charu Jain, Divia T. Joy and Rashmi Gupta

2.1 INTRODUCTION

Machine learning (ML), for the need of introduction, is a subset of artificial intelligence (AI) – another fancy of many technical researchers. Being a subset of AI, ML has achieved a major part of the implementation aspect of AI. In the perspective of ML researchers, Naqa and Murphy, a developing subfield of computing algorithms called "machine learning" aims to mimic human intelligence by taking in information from its surroundings [1]. Now that we have a fair idea of ML with respect to AI, let us take a peek into what more ML offers. ML encompasses a wide-ranging area of multidisciplinary facets belonging to a plethora of arenas that may or may not have to interact with computer and technical sciences.

Furthermore, having become ingrained in most scientific disciplines, more so in recent years, a plethora of modelling tools and even algorithms for a wide range of data processing tasks are now a part of the ML process [2]. Not only data processing, life sciences and even physical sciences make use of ML to regulate their theories and simulate subject-specific hypotheses. Moreover, electronics and digital sciences find ML indispensable in a developing environment that caters to the fact that ML is mostly software that can be integrated with hardware. Material classification and property definition are easier examples of ML implementations in this sphere. In their research, Kudelina and Vaimann have quoted ML for electrical sciences, most aptly, that AI algorithms can be trained using mathematical models of electrical machines. This is accurate given that the industry and laboratories have difficulties in gathering large amounts of data due to resource constraints [3]. Thus, several intriguing AI-based diagnostic methods are showcased with an emphasis on their characteristics.

With regard to core machine learning, we will first address its concepts and the theoretical structure that has riveted the world of AI at large, prior to which, we will define it independent of its application and working. Once the groundwork is laid, this chapter will address advanced ML, bearing in mind that ML itself is advanced in its fundamental structure. Finally, the chapter will conclude with the applications and a massive array of its relevance in electrical sciences and why it is a dazzling idea to integrate ML and semiconducting material. So let us dive right into the deep end!

DOI: 10.1201/9781003508304-2

Foremost, let us understand that AI is about imbuing devices with lifelike realism. It is interesting to note this because AI is all about causing machines to mimic humans [4]. Here is where we come to ML. It is that branch of AI which helps a machine to train itself on some data and find patterns within that so as to modify its analysis and apply it to the future in terms of analysis and prediction. The motive of ML is to interpret data and understand patterns. Humans look at data, and the brain, on its own accord, begins to analyse the data and make interpretations and inferences. For instance, a human baby sees its surroundings and finds the task that is most repetitive and stores that action and begins to use it for future situations: say a child asks for chocolate and realizes that if it is refused, he or she can get it by throwing a tantrum. That is exactly what machines do: they find data and learn from it, and then based on future inputs, they rely on the learning to perform event-specific actions to get the desired outcome. It is fascinating to watch the task! Figure 2.1 depicts AI as the superset of learning.

We now address the next task: ML concepts. The first implication is for data, which can be structured, unstructured, or semi-structured. Based on the data, we set up "tasks" for machines to gain "experience", and then we measure its "performance". This leads to ML categorization in terms of data input as supervised, unsupervised, semi-supervised, and reinforcement learning. All the four subsets have their own set of algorithms that define their operation. Next, we head to the applications of these intelligent systems. Xue and Zhu spearheaded research into fields that are affected by ML, and they concluded that ML is a basic technique that gives computers intelligence. Its application, which formerly relied mostly on methods of synthesis and induction rather than deduction, has already expanded to other areas of AI and all

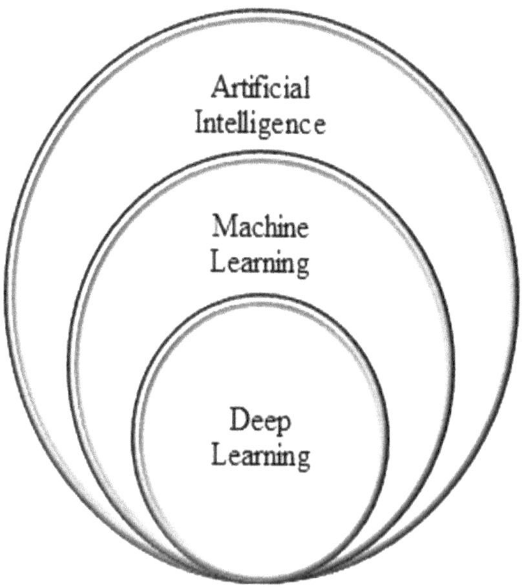

FIGURE 2.1 Artificial intelligence as the superset of learning.

known sciences with tremendous ease [5]. Now that we have laid the groundwork, with a basic idea of ML, let us walk towards the deeper end.

2.2 THE ORIGINS

AI, while formally introduced first in the 1950s, had its foundations built much earlier. From a poet's theory to research papers that questioned the idea, AI riveted early researchers even before being formally coined. ML is a crucial instrument for achieving the objective of utilizing AI-related technologies. It is sometimes confused with AI due to its capacity for learning and making decisions, but it is actually a subset. It was only after the 1970s that it split off to develop independently. A brief discussion on the origins of ML follows.

2.2.1 Artificial Intelligence

From the BC era to the 21st century, ML has had multiple phases. The term "artificial intelligence" originated, ironically, from the Greek word *automaton*, which means "behaving on one's own accord". The earliest records of automaton come from 400 BC, which is hardly surprising, as Greek scholars in the past were known for their above and beyond thoughts and philosophies. It was much later in the early 1900s when many ideas centred around artificial humans instigated scientists to question if it was possible to create an artificial brain that could replicate the human cycle of thoughts and behavioural patterns. The scholarly paper "A Logical Calculus of the Ideas Immanent in Nervous Activity" by Walter Pitts and Warren McCulloch from 1943 contained the first mathematical model of neural networks and therefore launched the field of ML [6].

The book *Organization of Behaviour*, by Donald Hebb, contained theories on the connection between brain activity, neural networks, and behaviour. First published in 1949, it became one of the foundational works in the history of ML. As mentioned earlier, AI came into being in the 1950s. From Alan Turing's seminal research paper "Computing Machinery and Intelligence" opening with the line, 'I propose to consider the question, "Can machines think?"' [7], to Arthur Samuel developing the preliminary learning process that could be applied to machines in 1952 and finally John MaCarthy coining the term "artificial intelligence", the world has been abuzz about how machines can be imbued with information that can be interpreted only by the human brain. Figure 2.2 highlights the timeline of AI to ML conversion.

2.2.2 Neural Networks

In 1957 Frank Rosenblatt gave the lowdown for the earliest neural grid for machines – the perceptron; this system was able to actually replicate the *Homo sapiens* brain in terms of processing thoughts. In the 1970s ML came up to speed, branching out as a subset of AI which helped with pattern analysis and interpretation for machines to understand and interpret data. Further, in the 1990s, ML, which had been running on derived knowledge, began to take the data-driven approach.

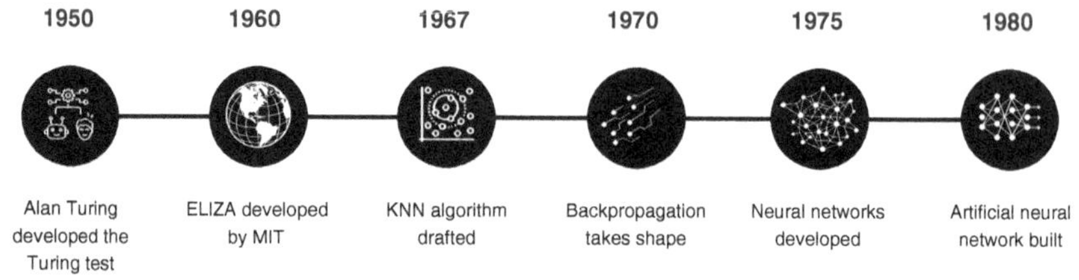

FIGURE 2.2 The AI to ML timeline.

In 1997 IBM's Deep Blue defeated the world champion at chess, furthering the idea of ML developing into deep learning, which was a term coined by Geoffrey Hinton. Since then, multiple multinational corporates began leveraging this knowledge engineering and deploying it to building models that could imitate human emotions and intelligence at large. Today ML has taken the world by storm with several use cases and applications beyond what could have been thought by the founding fathers of AI. Furthermore, research after research has determined that ML still has a long way to go and is set to stay.

ML powers a wide range of jobs across several industries, from finance experts seeking signals for profitable trades to data security organizations searching for viruses. AI algorithms are designed to learn continuously, simulating a virtual personal assistant, a function they excel at.

2.2.3 Ideating Machine Learning

The story of ML, once defined by the 1970s, led to its specifications. After the initial research into ML, the thought of neural networks took stage, which is essentially the principle of intelligent algorithms. Neural networks are framed to replicate the human brain. As it is a network of connected pieces, it is known as a neural network. Biological nerve systems studies serve as the basis for these components. Put another way, neural networks are an attempt to develop machines that, by utilizing components that behave like biological neurons, perform similarly to the human brain. A neural network's job is to take an input pattern and turn it into an output pattern. Pattern categorization is one of the tasks that a neural network may be trained to perform, and it will be thoroughly explained because this idea is somewhat abstract. The technique of classifying patterns into groups is known as pattern classification [8]. The basic building blocks for neural networks are similar to neurons. These components are linked to one another through connections that might change in strength according to an algorithm or learning process. To assess its level of activity, each of these units independently (in parallel) integrates the data from its synapses [9].

Neural networks led to the development of the perceptron. Resulting as a combination of Hebb's and Samuel's model, the perceptron was designed by Cornell Aeronautical Laboratory (CAL) to be a machine, not a program. The software was then installed and gave rise to the Mark 1 Perceptron instead of the IBM 704. As a result, the algorithms and software were portable and machine transferable. Later the subsets of ML prospects came into being with the development of the nearest neighbour algorithm, feedforward and backpropagation networks, boosting algorithms, and so on. The varieties of algorithms that came into being required a clear categorization under ML.

2.3 MACHINE LEARNING MODELLING

Before we dive into the types of ML, its categorizations, and the basis for the same, we will look at the parameters that define the grouping of ML and lay out the foundation of the various ML algorithms. The key elements of ML rest on three pillars, namely: representation, evaluation, and optimization [10]. The foremost deals with the what of an ML model's look and the how of knowledge representation. The second works around how good the model is, what its differentiators are, and the evaluation benchmarks for the algorithms involved. The last element is related to the process of seeking out a good model(s) and thus moving towards how the programs are actually generated.

2.3.1 Types of Machine Learning

On the basis of the aforementioned component declaration of ML, the concept itself can be broadly categorized into three types: supervised learning, unsupervised learning, and reinforcement learning [11]. Let it as well be made clear that as automated thinking techniques advanced, a fourth category was introduced within ML catalogues: semi-supervised learning [12]. We will now try to understand and differentiate the categories. Figure 2.3 shows the different ML subcategories.

2.3.1.1 Supervised Learning

As the name itself suggests, this model of learning has a supervisor, which we, in technical terms, refer to as "labels". Given a dataset with labels, an ML model can be constructed to train itself on the output, and then when given new, fresh data, the expectation remains to be able to retrieve the correct label. Information from one neural network serves as an instructional signal to change the connection pattern in another network through supervised learning. Since the advent of ML, supervised learning has developed under the concept that human behavioural patterns are formed in infancy based on data – for example, a child is shown a tree and repeatedly told, "This is a TREE"; then perhaps the child is shown a car and again is repeatedly told, "This is a CAR"; and so on with any number of unknowns that get categorized in the brain under a subheading. Essentially, this becomes supervised learning; much later, if the child is shown a picture that contains both a tree and a car, he or she will be able to identify them, not based on characteristics but based on the labels assigned to them. An ML model behaves more or less similarly. Figure 2.4 illustrates the supervised learning model.

As a result, the taught network picks up new skills in information processing that enable it to accomplish the specific objective or transformation that the instructional signal specifies. By doing this, supervised learning effectively and precisely creates connection patterns that are not required to be encoded and frequently cannot be.

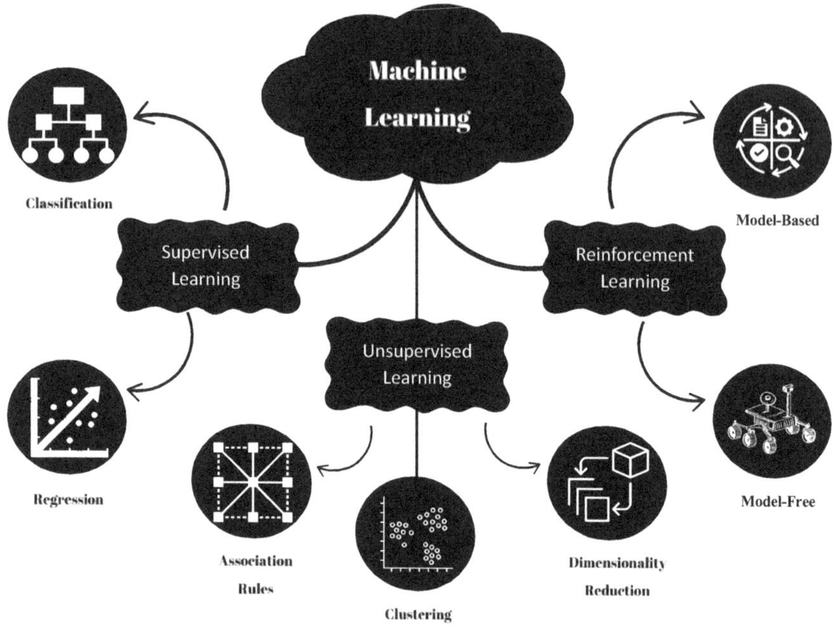

FIGURE 2.3 Machine learning subdivisions.

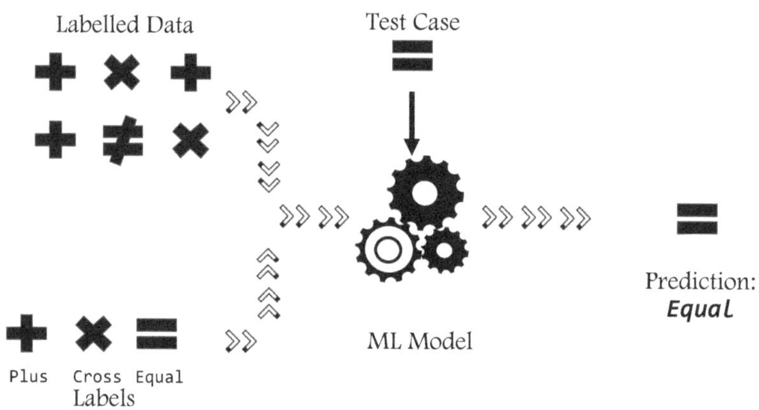

FIGURE 2.4 Supervised learning model.

Numerous neural activities are aided in the creation and upkeep of supervised learning [13]. Supervised learning algorithms are further categorized into classification algorithms and regression algorithms.

2.3.1.2 Unsupervised Learning

Fairly the opposite of supervised learning, unsupervised learning stems from the idea that instead of labels being formally available, the ML model must learn to identify

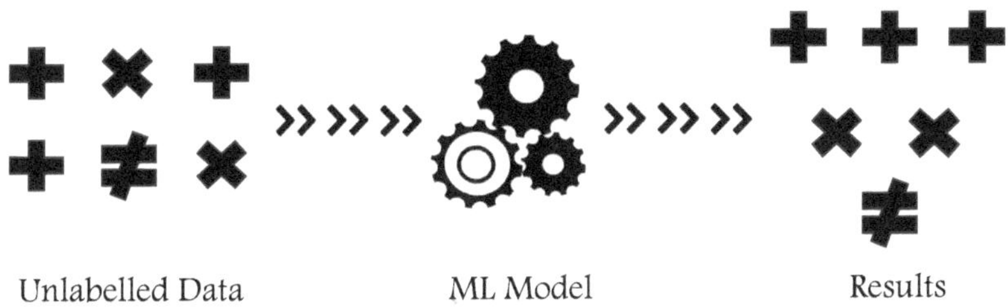

FIGURE 2.5 Unsupervised learning logical model.

and interpret similar features of the elements in a given dataset and then group them based on those similarities. The argument put forth is that one, the knowledge incorporated into maps or models is provided by the repetition of sensory messages. Then a portion of this information can be discovered by looking at the mean, variance, and covariance of sensory messages. Let us understand this using the child analogy. A child is given the cut-outs of five mathematical figures, two sets each. Say two squares, two circles, two triangles, and so on and then is asked to pair them up. The child is able to do so by learning the structure and characteristics of the object. Say, at a later time, a new square piece is cut out and given to the child; he or she will be able to place it with the other squares. Unsupervised learning works similarly. The algorithm can understand characteristics and similar patterns, so it knows WHERE something goes but does not know WHAT something is. Figure 2.5 depicts the unsupervised learning logical model.

Further, by using this information to build a model of "what usually happens", incoming communications may be automatically compared to it, making it possible to spot unexpected differences right away. Finally, the kind of knowledge that goes into such a filter is a required precondition for regular learning, and associations with the logical functions of the components – rather than just the elements themselves – can be formed with a representation of constituents that are independent [14]. Unsupervised learning algorithms are further classified as clustering, association rules, and dimensionality reduction algorithms.

2.3.1.3 Reinforcement Learning

This method of learning is most closely related to the technique that parents use in guiding and teaching children. Reinforcement learning mimics a trial-and-error method where success results in rewards and failure leads to punishment. Again, let us understand this with the child analogy. A few days after understanding trees, cars, and shapes, the child is shown a tree; for argument's sake, let us say that the child shouts, "That is a CAR"; the child's guardian will either scold or give a disapproving look and correct him or her. Now had the child said, "That is a TREE", he or she would have gotten an appreciation or a candy. This is almost exactly what reinforcement learning does.

To tackle reinforcement learning challenges, two primary approaches can be pursued. The first would be to investigate through the span of behaviours and identify something that works well in the given setting. This is precisely what genetic programming and algorithms have adopted as a strategy. The second is to calculate the

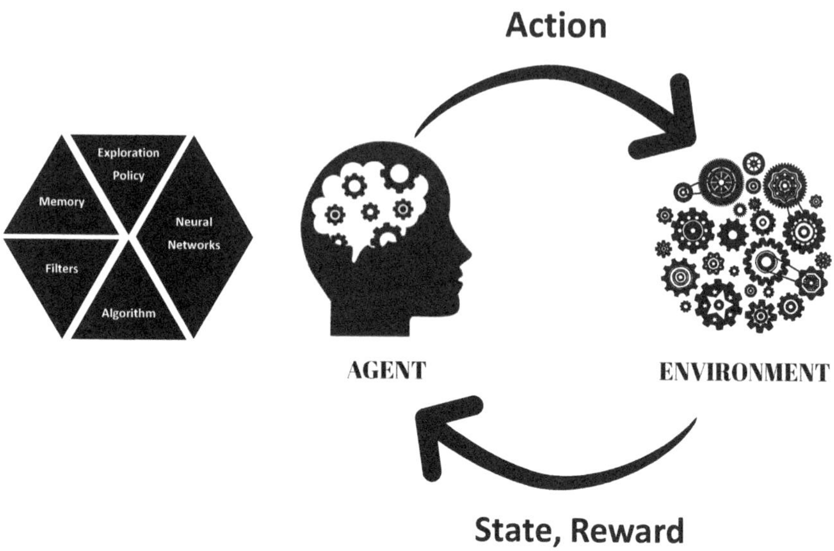

FIGURE 2.6 Representation of reinforcement learning.

utility of acting in the various conditions of the world using statistical and dynamic programming approaches [15]. Reinforcement learning algorithms are grouped into model-based and model-free algorithms, also referenced as positive and negative techniques. Figure 2.6 represents reinforcement learning.

2.3.1.4 Semi-Supervised Learning

Semi-supervised (SS) learning has a minimal dataset of labelled data and a vast set of unlabelled data of the same sort. When the ML model works with the determined combination of data and translates the features of labelled data to unlabelled data and then predicts the category of a fresh set of data, this results in SS learning. Here is the final child analogy: this child is now extra smart. He or she is shown only one or two trees and told, "This is a TREE". The child is taken on a walk the next day, where every standing tall structure is brown at the bottom and green at the top; now the child is able to stop and say, "TREE". This would be SS learning. Figure 2.7 summarizes a sample of semi-supervised learning.

SS learning, which involves concurrently learning a set of disjoint concepts, with the learning algorithm only having access to partial concept membership information, has now been added to the distribution-independent paradigm of (supervised) concept learning. Numerous learnable concept classes have been demonstrated to be SS-learnable as well. An intermediate oracle is used to introduce a new learning method. To be SS-learnable, enough parameters are provided for a set of concept classes [16]. SS learning algorithms can be catalogued as self-training, co-training, and graph-based label propagation.

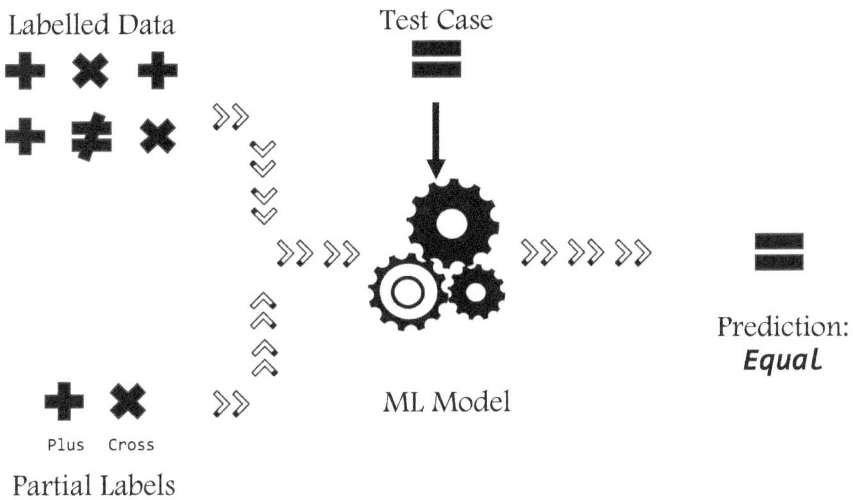

FIGURE 2.7 Sample of semi-supervised learning.

2.4 IMPLEMENTATION OF ML MODELS

We have taken a thorough look at the types and paraments of ML models. There is a clear definition in the use case design of the models. This section attempts to discuss the implementations of supervised and unsupervised algorithms on the UCI SECOM dataset. According to McCann and Johnston, a sophisticated, contemporary semiconductor manufacturing process is typically continuously observed through signal tracking or variables obtained from sensors or process measurement sites [17]. In their research, they developed a dataset that holds over 500 varieties of data signals that help extract the particularities of semiconductor development. The specific monitoring system technically measures the various signals and finally leads to a pass or fail in a contained house line testing. The dataset dimensions are given in Table 2.1.

The preprocessing of the dataset involved altering the representation and replacing *NaN* values with the mean/median based on data availability. The passes originally were represented by −1 and later converted to 1 during the data cleaning process; likewise, the fails, then represented by 1 were replaced with 0 to ensure the ease of apt classification activation functions. Excess columns were dropped, as the dataset is already normalized, and the empty or *NaN* values are replaced with their column-wise mean. The columns with missing values that go beyond the acceptable three standard division distances were dropped. Further, the columns with univariate values were done away with. The data once divided into parameters and targets was essentially split into partial datasets, one meant for training and another for testing. The training parameters were then scaled using *sklearn.preprocessing.MinMaxScaler*. Once the data was fit into the scaler, the final step of

TABLE 2.1
Original Dimensions

Features	Test Cases	Passes	Fails
592	1,567	1,463	104

TABLE 2.2
Modified Dimensions

Parameters	Target	Training	Testing
443	2	1,175 (75%)	392 (25%)

transformation was performed before the data needed to be run through the various models. The dimensions of the dataset as a result of the preprocessing are given in Table 2.2.

The dataset once processed underwent six major supervised learning algorithms, namely: linear regression, logistic regression (LR), decision tree, random forest, support vector machines, and *K*-nearest neighbours. This was followed by eliminating the labels and running the dataset under three major unsupervised algorithms. Each of the algorithms have been explained and demonstrated in the implementation.

2.4.1 Supervised Learning Algorithms

As mentioned earlier in the chapter, supervised learning involves a clarity in data labels. Here we further dissect supervised learning into classification and regression. The former involves algorithms that look for methods that can actually help categorize the dataset according to the subset factors affecting it. Applying a classification approach involves grouping the data according to the knowledge that the computer program has gained from running repeatedly through the training dataset. Regression, on the other hand, undertakes the task of finding correlations between independent and dependent variables. Some of the classification algorithms are k-nearest neighbour, naive bayes, random forest, LR, etc. And some of the regression algorithms include decision tree regression, principal component, support vector regression, linear regression, and so on. Before we dive into the algorithms, let us understand that all systems will use a confusion matrix as the final deciding element for accuracy for the classification models. Four parts make up a confusion matrix, and they could stand for the following: true positives, which are actual truths that are predicted as true; false negatives, when actual truths are predicted as false; true negatives, where actually false vales are predicted as false too; false positives, when actually false values get predicted as true. The distinction between true positives and true negatives is what is intended to increase so as to influence the accuracy [18]. Therefore, the accuracy is calculated as follows:

$$\frac{TP + TN}{TP + FP + FN + TN} \tag{2.1}$$

Further, for the regression models, we will be using the mean absolute error (MAE) and mean squared error (MSE) determinants. The two values are calculated as follows:

$$MAE = \frac{1}{n}\sum_{i=1}^{n}\left|y_i - \hat{y}_i\right| \tag{2.2}$$

$$MSE = \frac{1}{n}\sum_{i=1}^{n}\left(y_i - \hat{y}_i\right)^2 \tag{2.3}$$

In Equations 2.2 and 2.3, n is the total number of instances, y_i is the predicted value of instance i, and $\hat{y}_i$ is the real target value of i. The lower the value of MAE and MSE, the better the model [19].

Within all the supervised learning and unsupervised learning algorithms alike, the dataset variation over testing and training, the ratio of balance, and the type of scaling affect the accuracy of the data. For instance, within the classification LR, *StandardScaler* gives an accuracy of 0.82, while the *MinMaxScaler* raises the accuracy to 0.93, which is quite the accuracy jump, given the use of the same algorithm.

2.4.1.1 Linear Regression

A fundamental statistical method for figuring out the relationship between two continuous variables is linear regression. It assumes that there is a straight line between the independent and dependent variables. An unknown value can be predicted using linear regression analysis based on an analogous unknown value [20]. The variable that a researcher hopes to predict is known as the dependent variable. The variable that is used to predict the value of the other variable is known as the independent variable. Figure 2.8 illustrates the accuracy for linear regression.

In Figure 2.8, two terms help determine the accuracy; MAE and MSE are the accuracy calculators for regression models. The closer the two values are to 0, the

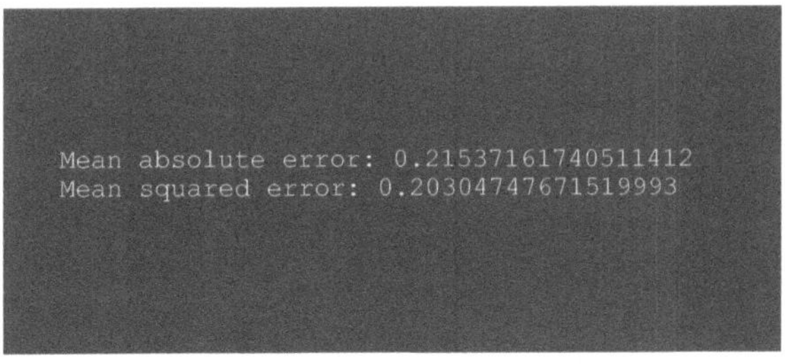

FIGURE 2.8 Accuracy for linear regression.

better the model. The main goal of LR is in finding the line that fits the data in the best possible way or, in the case of many predictors, where the hyperplane is involved. The sum of squared residuals, or errors, between the observed and predicted values is minimized to achieve this.

2.4.1.2 Logistic Regression

Logistic regression is a statistical model that allows for the division of data into discrete groups by looking at the correlation between one or more independent variables [21]. Predictive modelling, in which the model determines the mathematical probability of an occurrence falling into a specific category or not, frequently uses it. Logistic regression is applied to binary classification tasks in which there are two possible outcomes for a categorical outcome variable. It does this by fitting data to a logistic curve, which estimates the likelihood that an event will occur. The foundation of LR is the assumption of a linear relationship between the predictors and the outcome variable's log odds. Figure 2.9 explains the confusion matrix for LR.

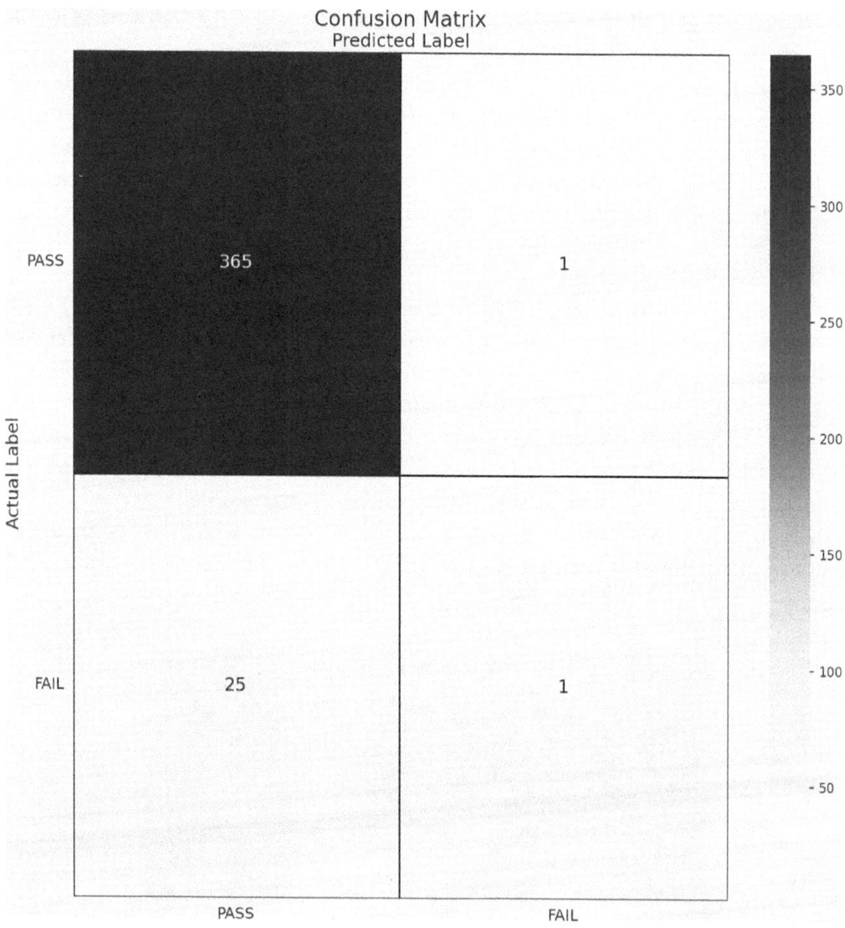

FIGURE 2.9 Confusion matrix for logistic regression.

According to the confusion matrix given in Figure 2.9, the 365 true positives give a clear idea of the efficiency of the model prediction, which in actuality produces an accuracy of 0.934, so additionally, it makes the assumption that each observation stands alone.

2.4.1.3 Decision Tree Regressor

Now the decision tree regressor (DTR) is essentially a supervised learning technique which is nonparametric in nature and used for regression primarily, but it also accommodates classification. The objective is to build a model that, by utilizing basic decision rules deduced from the data features, predicts the value of a target variable [22]. The constant approximation of a broken subset can be assumed as donning the structure of a tree. Now DTR makes relevant predictions on future data by using an object's qualities to train a model vis-à-vis the structure of a tree. And this reiterated process produces meaningful continuous output. Continuous output denotes a result or output that is not discrete – that is, not solely represented by a known, discrete collection of numbers or values. Figure 2.10 describes the accuracy for the decision tree classifier.

In Figure 2.10, the MAE and MSE help evaluate the model's accuracy, which is comparatively a good one, due to other model parameter tunings. Fine-tuning a model essentially involves the alteration of values given to the parameters, such as *random_state*, *test_size*, *criterion*, *max_depth*, etc.

2.4.1.4 Random Forest Classifier

Random forests are very useful for managing complicated and sizable datasets and high-dimensional feature spaces and offering insights into the significance of individual features. It can reduce overfitting and retain a high level of predicted accuracy [23]. Figure 2.11 explains the confusion matrix for the random forest classifier.

Accuracy rises up to 0.934 with 513 true positive values, which is comparatively higher than the previous classification and regression algorithm models.

2.4.1.5 Support Vector Machines

The primary goal in the support vector machine (SVM) algorithm is to detect the best hyperplane in all but an N-dimensional space. This n-D space then may be used for

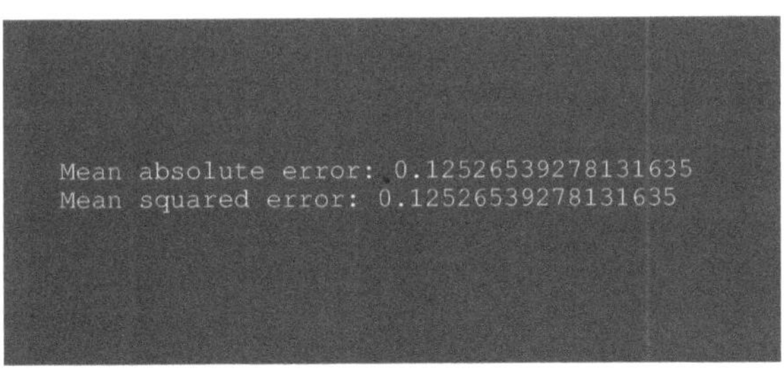

FIGURE 2.10 Accuracy for decision tree classifier.

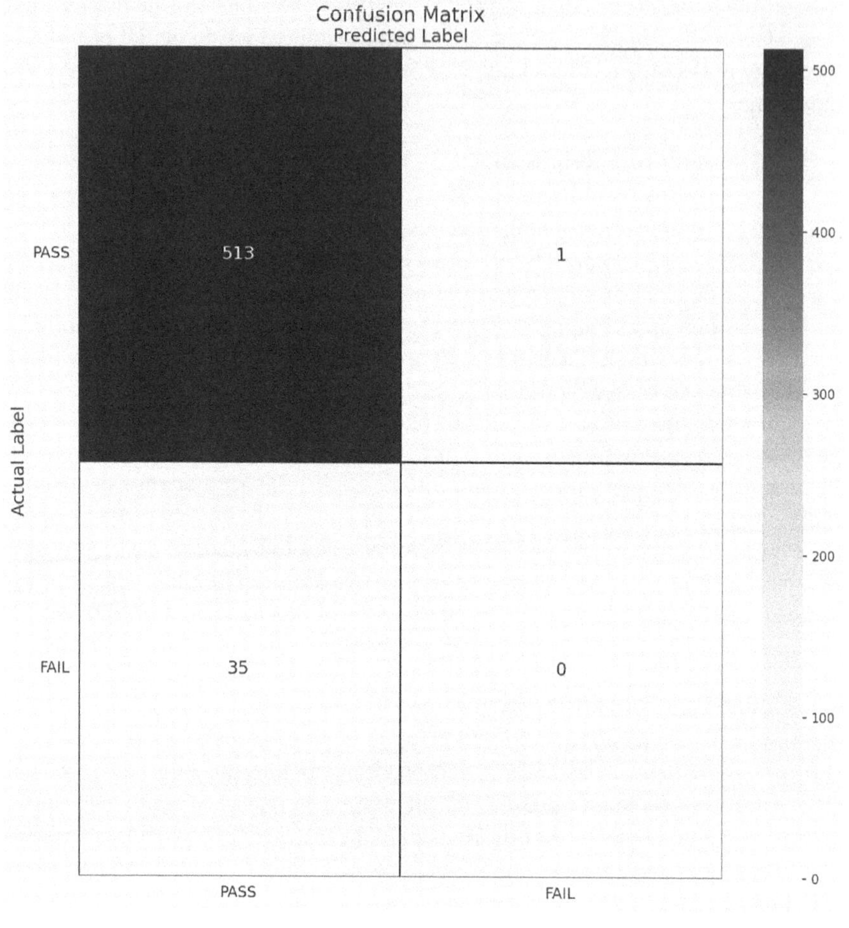

FIGURE 2.11 Confusion matrix for random forest classifier.

the purpose of dividing data points into multifaceted feature space groups [24]. The hyperplane makes certain efforts to preserve a buffer that needs to be large enough to accommodate antithetical distances between the points that are the closest within various subsets of divided classes. Figure 2.12 depicts the confusion matrix for SVM.

The classifier gives a peaking accuracy of 94.64% subject to similar data distributions along with the tuning parameter for the SVM model.

2.4.1.6 *K*-Nearest Neighbours

The design of the *K*-nearest neighbours (KNN) algorithm is founded on a basic distance mapping to locate a set number of closest neighbours, *K*. Using any predefined distance metric, such as the Euclidean distance or even the Manhattan distance, is a common approach in defining the vicinity of a given data point. The advantageous

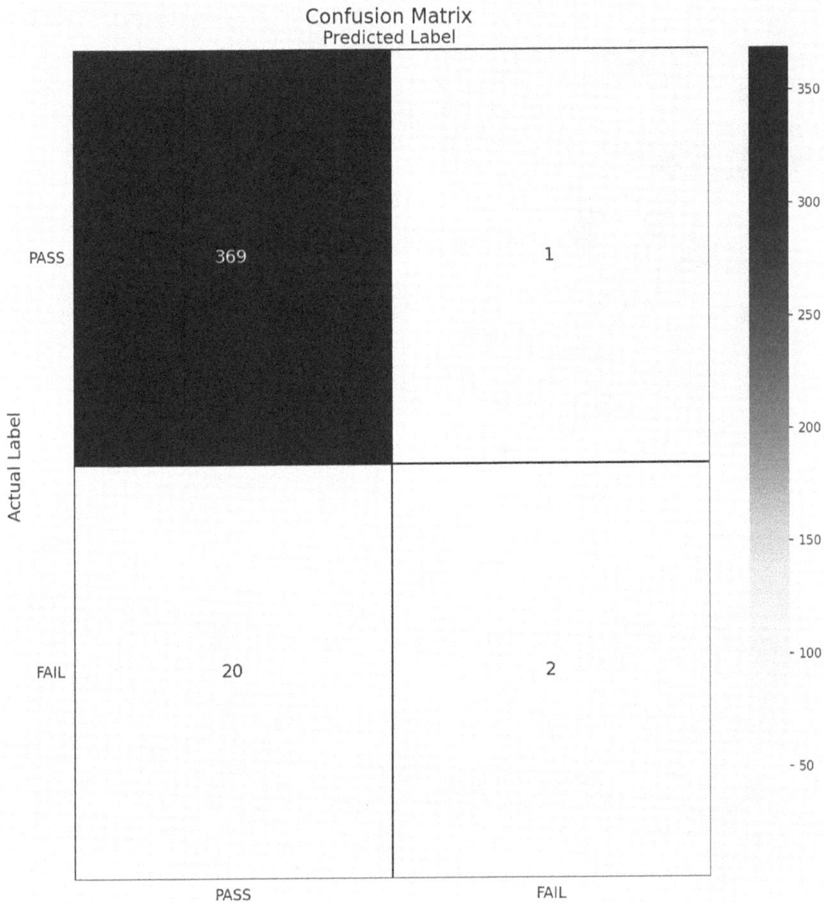

FIGURE 2.12 Confusion matrix for support vector machine.

elect or the most commonly closest of the *K* neighbours are then used to decide and assign the class or, in some cases, even the value of the data element [25]. Using this method enables the set of rules followed to anticipate outcomes based on the local formation of the data and adjust to various precedents. Figure 2.13 highlights the confusion matrix for KNN.

The matrix in Figure 2.13 produces an accuracy of 0.938 with *NULL* values for false positives and true negatives. Next, we take a look at the actual data distribution in the *K*-neighbour classifier. In Figure 2.14, the distance mapping of the data points has been done using Euclidean distance mapping. The score scale ranges from 0 to 1, and the intensity of the colour indicates the confidence of the model for assigning that score. Furthermore, the shapes denote the true labels from the dataset. Figure 2.14 shows the KNN map for predictions by the model.

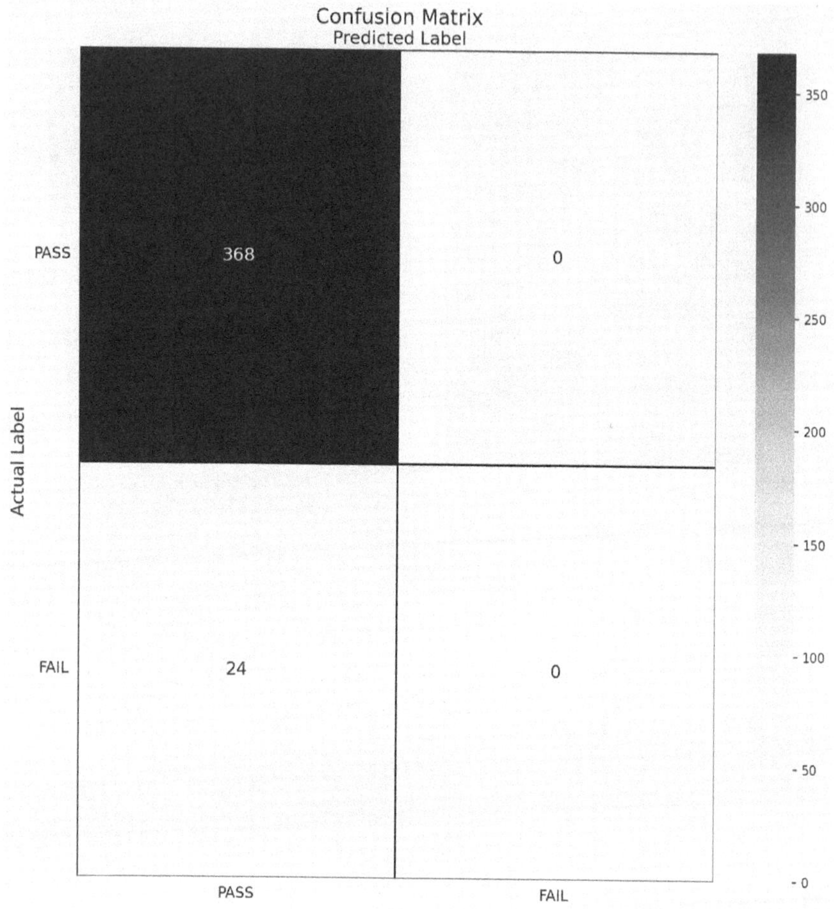

FIGURE 2.13 Confusion matrix for *K*-nearest neighbours.

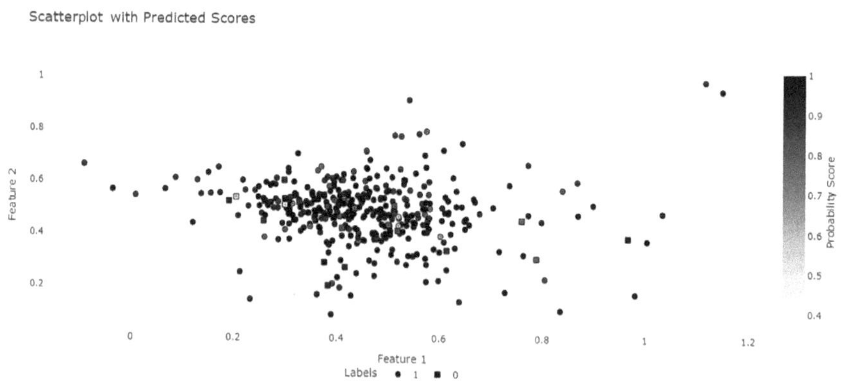

FIGURE 2.14 KNN map for predictions by the model.

2.4.2 Unsupervised Learning Algorithms

Once again, as mentioned earlier in the chapter, unsupervised learning involves a lack of data labels for the observations in the dataset; the resultant is a common clause grouping. Unsupervised learning too can be subdivided into three categories: clustering, association rules, and dimensionality reduction. For clustering, unlabelled data can be examined and divided into groups according to similarities or differences. The uncategorized data is divided into naturally occurring groups by identifying recurring patterns or similar structures. Second, the rule-based method called association rule mining can be used to find intriguing connections between data pieces in big databases. In unsupervised learning algorithms, correlations and co-occurrences within the data as well as the various connections between data objects are found by searching for frequent if-then linkages, also known as rules. Finally, dimensionality reduction is a method of unsupervised learning that lowers a dataset's feature count or dimensions. It reduces the amount of random or unnecessary characteristics in the dataset by extracting key features.

2.4.2.1 *K*-Means Clustering

K-means clustering is an unsupervised learning approach in which unlabelled data is made up or segregated into several clusters. The *K* in *K*-means is to specify the number of predefined clusters that need to be formed during the process. It simply means that in case $K = 2$, it will result in two clusters; in case $K = 3$, it will be three clusters, and so on. This algorithm allows us to organize data into separate groups and has the advantage of nearly automatically finding the categories of groups by itself from the unlabelled dataset without a need for training [26]. Each cluster in this technique has a centroid, generally the mean of the data points, attached to it according to centroid theory. Combining the distance of each point with its respective cluster executes the basic objective of this algorithm. Figure 2.15 depicts *K*-means clustering.

Unfortunately, the accuracy is as low as 44% using the unsupervised learning algorithm. For the investigation of the ideal number of clusters, a technique that is very commonly is the elbow method. The WCSS value is used in this procedure. Inside cluster sum of squares, or WCSS, is a statistical measure that characterizes the total variation inside a cluster. We can use any of the precalculated methods for distance mapping, such as the Manhattan or Euclidean distance, to measure the distance between data points and any corresponding centroid.

The silhouette coefficient, also called the silhouette score *K*-means, measures how similar a data point is within a cluster or how cohesive it is in relation to other clusters, which is known as separation [27]. Compared to the former, silhouette provides greater precision.

2.4.2.2 Hierarchical Clustering

In this algorithm, we start by treating each individual data point as an independent cluster of its own; then these are combined in twos and threes and so on to finally form a single mother cluster. The dendrogram is the structure that results from developing the hierarchy of clusters in this algorithm in the shape of a tree. The datasets are grouped into clusters using a bottom-up methodology. This means that the

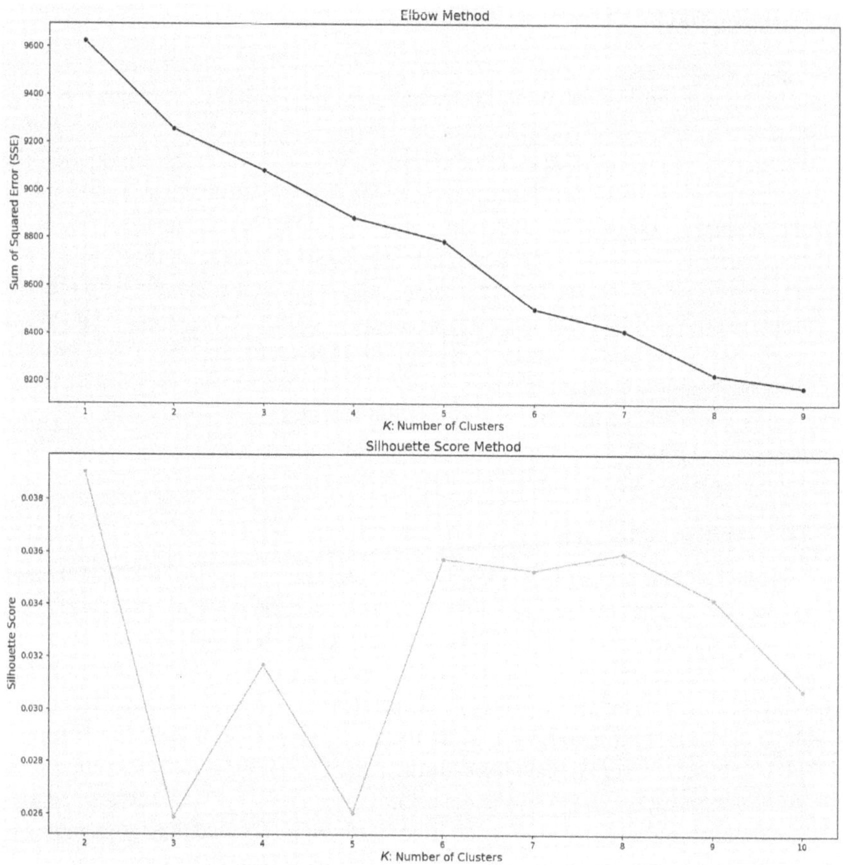

FIGURE 2.15 *K*-means clustering using the elbow method and the silhouette score method.

algorithm clusters those that are closest [28]. And to expand, this process is continued until every cluster is combined to form a single cluster that houses every dataset. Contrary to partition-based clustering, agglomerative clustering forms a sourced merge tree that is identified of binary characteristics as it moves from the nodes that are the least – that is, the leaves containing data pieces all the way to the root, which essentially holds the entire dataset. Figure 2.16 highlights the hierarchical clustering dendrogram and Figure 2.17 depicts the principal component analysis.

2.4.2.3 Principal Component Analysis

In principal component analysis (PCA), technically, the aim is dimensionality reduction following the creation of some new variables known as principal components; these originate by blending or merging the available original unknowns in a simple linear fashion. In essence, most of the facts included in the absolute variables are compressed or flattened into the preliminary components as a result of ex post facto combinations, thus giving some unknowns that are in no way correlated, and these

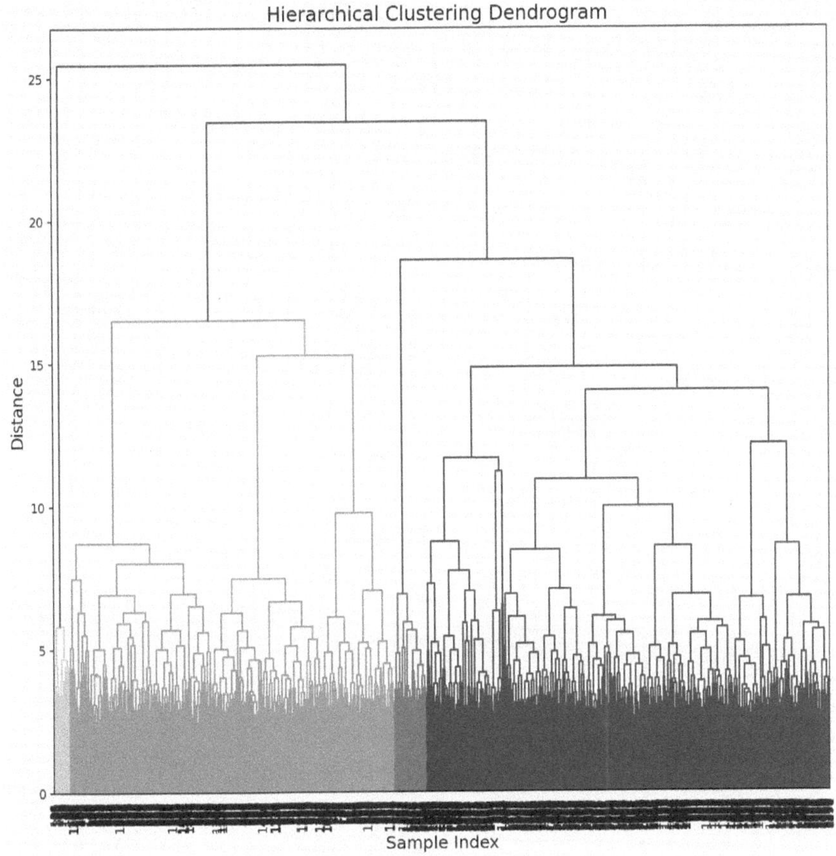

FIGURE 2.16 Hierarchical clustering dendrogram.

are called the principal components. The number of characteristics or dimensions in a dataset exponentially increases the amount of data required to get a statistically meaningful result. Overfitting, protracted computation delays, and a decline in the accuracy of ML models are the possible outcomes of this [29]. The "curse of dimensionality" refers to the problems that can arise while working with high-dimensional data. PCA is therefore necessary.

2.5 ANALYSIS OF ALGORITHMS

All the study so far has revealed much in terms of probable use cases in the field of electrical sciences, especially where semiconductor materials are concerned. Supervised learning methodologies and algorithms give much more reliable outcomes given that there is enough labelled data initially available for model training. Antithetical to this, unlabelled data leads to successful work towards feature extraction and analysis using unsupervised learning techniques. Figure 2.18 shows an outline of the data range.

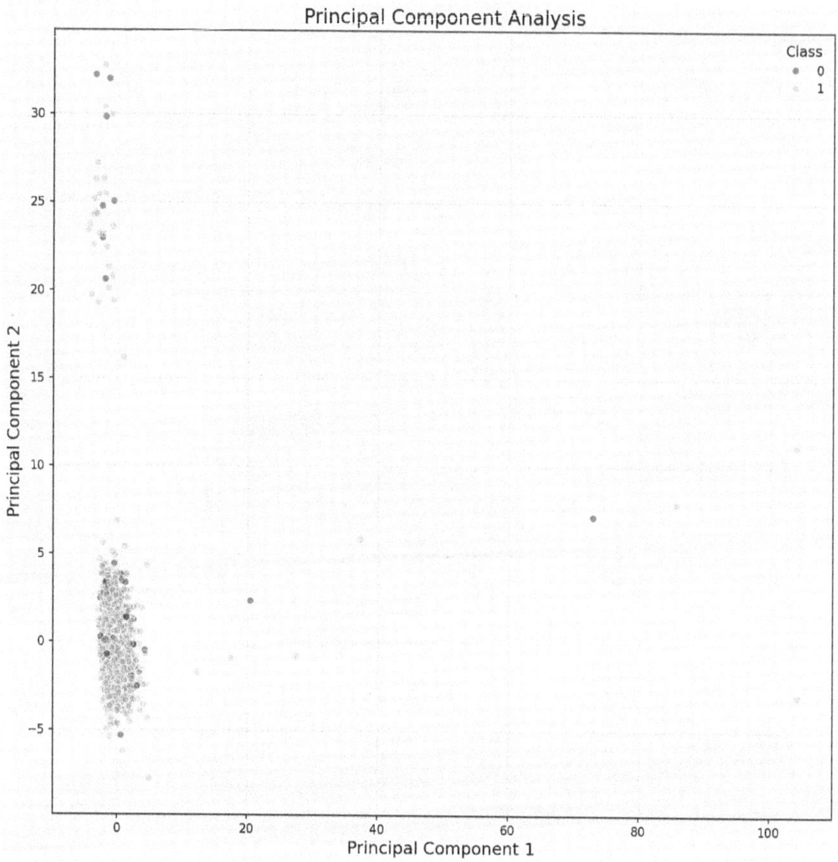

FIGURE 2.17 Principal component analysis.

In Figure 2.18, data fixations after each level of preprocessing have been defined. The paraments are subject to the data cleaning process applied column-wise, and the test cases change only as a result of data split, else the data length remains consistent. This essentially means that there is no data loss in the analysis of the signal collections. Variation refers to the alteration of data based on level-wise flow; it depicts the difference in the form and dimension of data from the previous preprocessing.

Supervised learning algorithms are a category of ML techniques where the model is trained on labelled data, meaning the input data has corresponding output labels. We have seen that linear regression is used to predict continuous numeric values. Logistic regression predicts the probability of a binary outcome. Decision trees help make decisions based on a set of if-else conditions derived from the training data. Random forest is an ensemble method that combines multiple decision trees to improve predictive accuracy and reduce overfitting. SVM finds the optimal hyperplane that separates classes in high-dimensional space. KNN predicts the value of a new data point based on the majority class of its *K*-nearest neighbours. Then there is

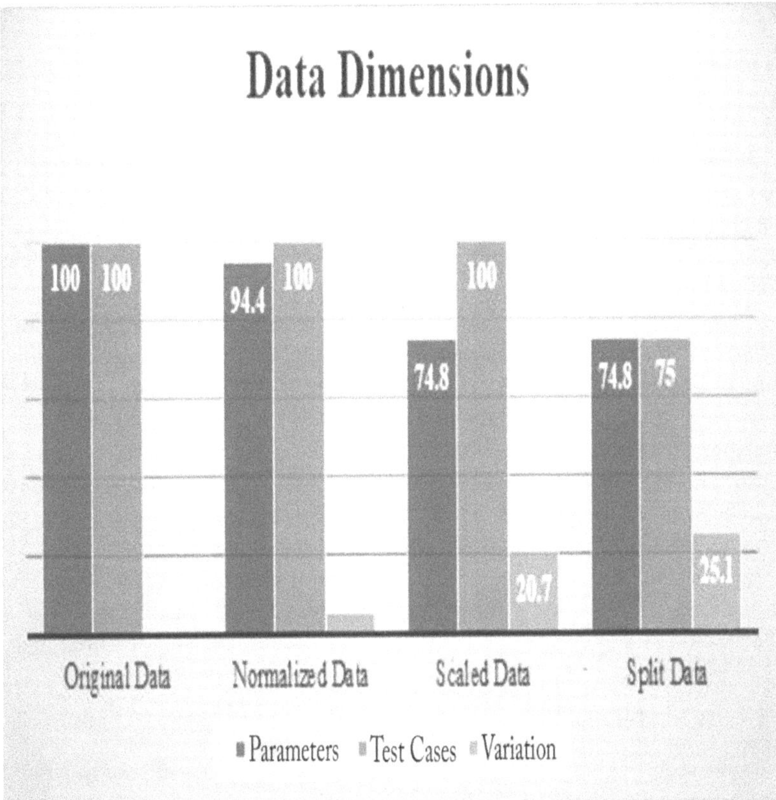

FIGURE 2.18 Dimensions of data over preprocessing.

naive bayes, which uses Bayes's theorem to predict the probability of a class given input features. Other than these, gradient boosting machines combine weak learners sequentially to improve predictive performance. Furthermore, neural networks learn complex patterns and relationships in data through layers of interconnected nodes (neurons).

Unsupervised learning algorithms are used to find patterns and structures in data where the data is not labelled or categorized. Here's a summary of some popular unsupervised learning algorithms: *K*-means divides data into clusters based on similarity. Hierarchical clustering starts with each point as its cluster and recursively merges clusters based on distance until all points belong to a single cluster. PCA reduces the dimensionality of data while preserving its variance. We also have *t*-distributed stochastic neighbour embedding; density-based spatial clustering of applications with noise (DBSCAN); and association rule learning algorithms such as a priori algorithm, which identifies frequent item sets in transactional data; then there are anomaly detection algorithms and generative models.

Finally, as we have seen, in the case of supervised learning algorithms, the accuracy scores are higher as compared to unsupervised learning algorithms.

2.6 MODELLING ML FOR ELECTRICAL SCIENCES

AI has revolutionized established approaches, optimized systems, and stimulated innovation in the electrical sciences. This represents a paradigm shift. Innovative applications such as intelligent sensors for condition monitoring, autonomous robots for maintenance duties, and ML algorithms for renewable energy forecasts are made possible by AI-driven research in the electrical sciences. These cutting-edge uses not only progress the profession but also help solve societal issues and promote economic expansion.

While ML can help humans become more intelligent in specific areas, including decision-making and pattern identification, its ability to greatly increase human intelligence in general is still constrained by inherent cognitive distinctions and moral considerations. Where electrical sciences are concerned, ML models can go a long way in assisting the development of efficient objects and devices that can garner a safer future where resource wastage is reduced while reuse and life span are increased.

Choosing a platform can be difficult since the incorrect one can increase expenses or restrict the use of other useful tools or technologies. There is frequently a propensity to believe that a system with more features is superior when evaluating several problem statements to choose an ML model. Perhaps, but first researchers should consider what the ML platform will do with their data and the resulting output. Which features are essential to achieving the necessary ML capabilities? The functionality of an entire system could be doomed by a single missing feature.

2.7 ETHICS AND SOCIETAL IMPACT

In the entire technical analysis of the core ML algorithm design, we must address this one aspect to aid further studies. It goes without saying that man fears what he cannot control. Where automated intelligence is concerned, even in the sphere of electrical sciences, the ethical considerations are as important as the impacts of ML on society. Let us quickly dive into the ethical implications; for starters, the bias and fairness of ML models are flighty, based on the data they are trained on. Then there is the concern of the ML black box and the accountability required to vindicate any algorithm. Moving to the core area addressed in this book, resource extraction and sustainability are fundamental to the success of ML for semiconductor materials. Finally, regarding the societal impact, keeping in mind the economic factor, one must ensure that research or subsequent experiments in this are thrive, but without thwarting growth, in terms of geopolitics, affordability, and regulatory frameworks. It is good to not lose sight of the fact that the motive of ML is to allay the digital divide and not exacerbate it.

REFERENCES

1. El Naqa, I., Murphy, M.J. (2015). What is Machine Learning? In: El Naqa, I., Li, R., Murphy, M. (eds.), *Machine Learning in Radiation Oncology*. Springer, Cham. https://doi.org/10.1007/978-3-319-18305-3_1.
2. Carleo, G., Cirac, I., Cranmer, K., Daudet, L., Schuld, M., Tishby, N., Vogt-Maranto, L., Zdeborova, L. (2019). Machine Learning and the Physical Sciences. *Reviews of Modern Physics*, 91(4). American Physical Society. https://doi.org/10.1103/RevModPhys.91.045002.

3. Kudelina, K., Vaimann, T., Asad, B., Rassõlkin, A., Kallaste, A., Demidova, G. (2021). Trends and Challenges in Intelligent Condition Monitoring of Electrical Machines Using Machine Learning. *Applied Sciences*, 11(6), 2761. https://doi.org/10.3390/app11062761.
4. Xu, Y., Liu, X., Cao, X., Huang, C., Liu, E., Qian, S., Liu, X., Wu, Y., Dong, F., Qiu, C-W., Qiu, J., Hua, K., Su, W., Wu, J., Xu, H., Han, Yong, F., Chenguang, Y., Zhigang, L, Miao, R., Ronald, D., Sabine, V., Marko, K., Fredrick, Z., Ze, Z., Lifu, Z., Taolan, D., Ji, Y., Jialiang, L., Liang, L., Ming, L., Zhaofeng, A., Tao, Z., Bin, H., Xiao, C., Shan, L., Xiaohong, Z., Wei, L., James P., Tiedje, James M., Wang, Q., An, Z., Wang, F., Zhang, L., Huang, T., Lu, C., Cai, Z., Wang, F., Zhang, J. (2021). Artificial Intelligence: A Powerful Paradigm for Scientific Research. *The Innovation*, 2(4), 100179. ISSN 2666-6758, https://doi.org/10.1016/j.xinn.2021.100179.
5. Xue, M., Zhu, C. (2009). A Study and Application on Machine Learning of Artificial Intell-ligence. *2009 International Joint Conference on Artificial Intelligence*, Hainan, China, pp. 272–274. https://doi.org/10.1109/JCAI.2009.55.
6. McCulloch, W.S., Pitts, W. (1943). A Logical Calculus of the Ideas Immanent in Nervous Activity. *Bulletin of Mathematical Biophysics*, 5, 115–133. https://doi.org/10.1007/BF02478259.
7. Turing, A.M. (2009). Computing Machinery and Intelligence. In: Epstein, R., Roberts, G., Beber, G. (eds.), *Parsing the Turing Test*. Springer, Dordrecht. https://doi.org/10.1007/978-1-4020-6710-5_3.
8. Picton, P. (1994). What is a Neural Network? In: *Introduction to Neural Networks*. Palgrave, London. 978-0-333-61832-5 978-1-349-13530-1, https://doi.org/10.1007/978-1-349-13530-1_1.
9. Abdi, H. (1994). A Neural Network Primer. *Journal of Biological Systems*, 2. https://doi.org/10.1142/S0218339094000179.
10. Langley, P. (1996). *Elements of Machine Learning*. Morgan Kaufmann.
11. Shanahan, J.G. (2000). Machine Learning. In: *Soft Computing for Knowledge Discovery. The Springer International Series in Engineering and Computer Science* (Vol. 570). Springer, Boston, MA. https://doi.org/10.1007/978-1-4615-4335-0_7.
12. Board, R., Pitt, L. (1989). Semi-Supervised Learning. *Machine Learning*, 4, 41–65. https://doi.org/10.1007/BF00114803.
13. Knudsen, E.I. (1994). Supervised Learning in the Brain. *The Journal of Neuroscience*, 14(7), 3985.
14. Barlow, H.B. (1989). Unsupervised Learning. *Neural Computation*, 1(3), 295–311. https://doi.org/10.1162/neco.1989.1.3.295.
15. Kaelbling, L.P., Littman, M.L., Moore, A.W. (1996). Reinforcement Learning: A Survey. *Journal of Artificial Intelligence Research*, 4, 237–285. https://doi.org/10.1613/jair.301.
16. Board, R., Pitt, L. (1989). Semi-Supervised Learning. *Machine Learning*, 4, 41–65. https://doi.org/10.1007/BF00114803.
17. McCann, M., Johnston, A. (2008). *SECOM. UCI Machine Learning Repository*. https://doi.org/10.24432/C54305.
18. Ting, K.M. (2017). Confusion Matrix. In: Sammut, C., Webb, G.I. (eds.), *Encyclopedia of Machine Learning and Data Mining*. Springer, Boston, MA. https://doi.org/10.1007/978-1-4899-7687-1_50.
19. Pal, R., (2017). Chapter 4 – Validation Methodologies. In: Pal, Ranadip (ed.), *Predictive Modeling of Drug Sensitivity* (pp. 83–107). Academic Press. ISBN 9780128052747, https://doi.org/10.1016/B978-0-12-805274-7.00004-X.
20. Weisberg, S. (2005). *Applied Linear Regression* (Vol. 528). John Wiley & Sons.
21. Schein, A.I., Ungar, L.H. (2007). Active Learning for Logistic Regression: An Evaluation. *Machine Learning*, 68, 235–265. https://doi.org/10.1007/s10994-007-5019-5.

22. Kushwah, J. S., Kumar, A., Patel, S., Soni, R., Gawande, A., (2022). Comparative Study of Regressor and Classifier with Decision Tree Using Modern Tools. *Materials Today: Proceedings*, 56(Part 6), 3571–3576. ISSN 2214-7853, https://doi.org/10.1016/j.matpr.2021.11.635.
23. Liu, Y., Wang, Y., Zhang, J. (2012). New Machine Learning Algorithm: Random Forest. In: Liu, B., Ma, M., Chang, J. (eds.), *Information Computing and Applications. ICICA 2012. Lecture Notes in Computer Science* (Vol. 7473). Springer, Berlin, Heidelberg. https://doi.org/10.1007/978-3-642-34062-8_32.
24. Hearst, M.A., Dumais, S.T., Osuna, E., Platt, J., Scholkopf, B. (1998, July–August). Support Vector Machines. *IEEE Intelligent Systems and their Applications*, 13(4), 18–28. https://doi.org/10.1109/5254.708428.
25. Guo, G., Wang, H., Bell, D., Bi, Y., Greer, K. (2003). KNN Model-Based Approach in Classification. In: Meersman, R., Tari, Z., Schmidt, D.C. (eds.), *On the Move to Meaningful Internet Systems 2003: CoopIS, DOA, and ODBASE. OTM 2003. Lecture Notes in Computer Science* (Vol. 2888). Springer, Berlin, Heidelberg. https://doi.org/10.1007/978-3-540-39964-3_62.
26. Na, S., Xumin, L., Yong, G. (2010). Research on k-Means Clustering Algorithm: An Improved k-Means Clustering Algorithm. *2010 Third International Symposium on Intelligent Information Technology and Security Informatics*, Jian, China, pp. 63–67. https://doi.org/10.1109/IITSI.2010.74.
27. Shahapure, K.R., Nicholas, C. (2020). Cluster Quality Analysis Using Silhouette Score. *2020 IEEE 7th International Conference on Data Science and Advanced Analytics (DSAA)*, Sydney, NSW, Australia, pp. 747–748. https://doi.org/10.1109/DSAA49011.2020.00096.
28. Nielsen, F. (2016). Hierarchical Clustering. In: *Introduction to HPC with MPI for Data Science. Undergraduate Topics in Computer Science*. Springer, Cham. https://doi.org/10.1007/978-3-319-21903-5_8.
29. Abdi, H., Williams, L. J. (2010, July–August). Principal Component Analysis. *WIREs Computational Statistics*, 2(4), 433–459. https://doi.org/10.1002/wics.101.

3 Fault Detection in Semiconductor Manufacturing

A Classification Analysis of the SECOM Dataset

Vanshika Jain, Rashmi Gupta, Neeraj Gupta and Prashant Kumar

3.1 INTRODUCTION

In the cut-throat world of semiconductors, the efficiency and reliability of the manufacturing process are key factors. The production of semiconductors entails a sequence of complex procedures, each with the potential to introduce flaws that impact the yield or the percentage of products that meet quality benchmarks. Optimizing yield is vital since it directly affects the cost, quality, and overall profitability of semiconductor devices. Detecting and resolving problems during the manufacturing stage can assist in reducing waste, improving production efficiency, and averting the release of defective products.

The SECOM dataset has been thoroughly researched in the area of identifying and categorizing faults. It consists of data gathered from sensors used in the production process. Each record in the dataset captures a moment in time of different sensor measurements along with a tag indicating whether a yield defect is present or not. With over 1,500 instances and more than 590 attributes, this dataset presents challenges but is highly beneficial for creating and testing algorithms for defect detection. Examining this dataset can offer valuable insights into the causes of yield problems and assist in building predictive models for spotting faults at an early stage.

The focus of this section is to explore the SECOM dataset specifically for identifying yield defects through the use of classification algorithms. The main goals are to examine and preprocess the SECOM dataset, perform a thorough exploratory data analysis, assess and implement various classification algorithms, evaluate their effectiveness, and analyse the outcomes. Accomplishing these goals will highlight how sophisticated data analysis and machine learning methods can improve defect detection in semiconductor manufacturing processes, leading to better yield and operational efficiency.

DOI: 10.1201/9781003508304-3

3.2 IMPORTANCE OF YIELD OPTIMIZATION

In the semiconductor manufacturing sector, yield optimization plays a crucial role in determining the overall productivity, profitability, and cost-effectiveness of production operations. Here is a thorough analysis of its significance:

a. **Economic Effect:** Yield directly influences production expenses. Higher yield rates reduce the cost per unit, as a larger percentage of manufactured units meets quality standards. Conversely, lower yields lead to increased costs due to waste and rework. By optimizing yield, businesses can significantly boost their profit margins and gain a competitive edge in the market.
b. **Efficiency of Resources:** Yield optimization encourages the use of resources such as labour, materials, and energy. Fewer defects result in less material waste and reduced rework means conserving resources and minimizing the environmental impact of the manufacturing process.
c. **Ensuring Quality:** Yield optimization ensures a greater proportion of products meets quality and reliability standards. This is particularly critical in the semiconductor industry where product failures can significantly affect the performance and safety of electronic devices.
d. **Contentment of the Client:** Better yields are associated with more consistent and high-quality products, which increases customer satisfaction. Reliable goods promote long-term connections and increase consumer trust, which support ongoing corporate success.
e. **Advantage against Competition:** Companies that achieve yields in the fast-moving semiconductor sector can offer products at competitive prices without conceding on quality. This capability allows businesses to capture a market share, providing them with a significant competitive advantage.
f. **R&D and Innovation:** Yield optimization enables the reallocation of resources towards research and development (R&D) initiatives. By minimizing losses and inefficiencies, companies can channel more funds towards advancements in technology and processes that enhance their product offerings and strengthen their competitive position.
g. **Control of Risks:** High yields lessen the risk involved in production. When fault rates are lower, recalls and warranty claims, which can be costly and harm a company's reputation, are less frequent. Optimizing yields contributes to a more resilient and consistent production process.
h. **Stability of the Supply Chain:** Improving yields enhances the reliability of the supply chain. Steady production rates ensure fewer interruptions, allowing for better scheduling and communication with suppliers and clients. This consistency is crucial for maintaining supplier relationships and meeting delivery schedules.
i. **Adherence to Regulations:** Adhering to regulations is crucial in the semiconductor industry. Consistent quality is necessary to meet regulatory requirements and avoid issues. Yield optimization plays an important role in achieving this.

j. **Progress in Technology:** Leveraging technologies such as intelligence, machine learning, and data analytics is often essential for optimizing yields. These technologies not only boost yields but also drive overall technological advancement within the company, positioning it as an innovator in the field.

In short, maximizing yield in semiconductor production is vital for maintaining a stable supply chain, meeting regulatory standards, fostering innovation, ensuring cost-effectiveness, conserving resources, upholding quality control, enhancing customer satisfaction, gaining a competitive edge, and driving technological progress. By prioritizing yield optimization, companies can enhance their efficiency and position in the market significantly.

3.3 COMMON ISSUES AND FAULTS AFFECTING YIELD

The yield in semiconductor manufacturing can be significantly affected by common issues and mistakes, leading to increased production costs and decreased efficiency. One of the key challenges is the presence of impurities or irregularities in materials like silicon wafers, which can result in defects in the final products. For instance, impurities can cause variations in hardness that impact the overall quality of semiconductor devices. Another crucial factor is the consistency of the manufacturing process. Even slight fluctuations in temperature, pressure, or chemical concentrations can lead to irregularities and defects in semiconductor devices. These variations can occur at various stages of the manufacturing process, such as wafer production, wafer testing, assembly or packaging, and final testing, each presenting its own set of opportunities and challenges [1].

When machinery malfunctions, it can lead to faulty components or an inability to maintain the precise conditions necessary for high-quality production, resulting in a reduced output. The situation can be exacerbated by the complexity and cost of equipment like advanced lithography machines, making maintenance and replacement difficult. Even trace amounts of contamination from substances or particles in a cleanroom setting can adversely affect the quality of semiconductor devices. This contamination can occur during various manufacturing stages, such as doping, etching, and deposition, and can be challenging to detect and eliminate. If the design deviates from performance standards, defects in the semiconductor circuits or layout could lead to failures and impact productivity. These issues can be challenging to identify and rectify, requiring diagnostic approaches.

Defects in the production process can be caused by errors such as improper equipment setup or handling. To minimize this risk, operators can be trained rigorously, and quality control measures can be implemented to ensure adherence to standards and protocols. Moreover, insufficient testing and inspection procedures may overlook defects in a product at an early stage, allowing it to proceed further on the production line and reducing overall output. This highlights the importance of having robust quality control systems and advanced diagnostic techniques to minimize errors and optimize semiconductor manufacturing yield. Last, as production equipment ages and wears down over time, its performance can decline, impacting output quality and consistency. Regular maintenance upgrades and equipment replacements can address this issue by ensuring that machinery continues to operate with precise specifications [2].

To tackle these challenges in semiconductor production, stringent quality control protocols, continuous monitoring, and advanced diagnostic techniques are essential. Companies like Brewer Science are adopting approaches such as metrology and fingerprinting to reduce material flaws and contaminants, ensuring semiconductor devices with improved yield, performance, consistency, and reliability. Addressing these issues requires a strategy that encompasses quality control measures, regular oversight of manufacturing conditions, routine maintenance, and comprehensive testing procedures. Additionally, advanced diagnostic methods can be employed to identify and resolve issues affecting yield, boosting the efficiency and reliability of the semiconductor manufacturing process. Notable examples of these methods include monitoring systems and machine learning–driven fault detection.

3.4 BACKGROUND OF THE SECOM DATASET

A dataset that is commonly used for detecting and classifying defects in semiconductor manufacturing R&D is the SECOM dataset. This dataset is derived from a semiconductor manufacturing process where sensors collect several types of data related to the production of semiconductor devices. The SECOM project aims to improve knowledge and optimize manufacturing processes in the semiconductor industry to make this dataset accessible.

The dataset consists of readings obtained at different stages of the production process. These readings encompass numerous factors, such as pressure, temperature, chemical concentrations, and other variables. To provide an overview of the industrial setting, these measurements are taken at intervals.

In the structured data, each entry corresponds to a batch or production run of semiconductor devices. The dataset includes recorded sensor readings for each entry along with a label indicating whether a defect was detected in the yield for that particular batch. This label is crucial for training classification models, as it provides the target variable that algorithms aim to predict. To facilitate analysis, the dataset is typically divided into segments such as training and testing subsets. Machine learning models are developed and fine-tuned using the training set, and their performance on previously unseen data is evaluated using the testing set. This division allows researchers to assess how well their models generalize to new, untested datasets.

The SECOM dataset provides insights into the manufacturing process, capturing both routine and challenging scenarios. This makes it valuable for developing and accessing algorithms for fault detection. By leveraging this dataset, researchers can employ machine learning methods, such as anomaly detection and classification, to uncover patterns associated with issues and enhance the precision of their predictions. Overall, the SECOM dataset serves as a resource for refining defect detection methods in the semiconductor manufacturing sector. Its diverse range of features and real-world relevance make it a benchmark for evaluating various data analyses and machine learning techniques within this domain.

3.5 KEY TECHNIQUES AND ALGORITHMS IN FAULT DETECTION

Identifying faults plays a role in maintaining high yield rates and ensuring product quality in the semiconductor industry. Different approaches and algorithms are employed to detect and rectify errors in the production process. Here is a thorough examination of the various essential methods and widely applied algorithms for fault detection, along with an analysis of their efficacy:

3.5.1 Statistical Process Control

Statistical process control (SPC) is a crucial approach for leveraging methods to oversee and regulate manufacturing processes. SPC examines data collected from processes to identify variations and trends that may indicate potential issues. The tools commonly used in SPC include the following:

- **Control Charts:** These charts track the performance of a process over time and help identify any deviations from expected behaviour. They assist in determining whether differences are due to causes suggesting potential problems or common causes inherent to the process.
- **Pareto Analysis:** Based on the Pareto principle (80/20 rule), which asserts that 80% of issues are frequently caused by 20% of the components, this technique determines the most significant factors leading to errors [3].

Effectiveness: SPC is effective at monitoring processes and detecting deviations at any stage. However, it may face challenges in identifying subtle or complex faults that necessitate more advanced analysis methods.

3.6 MACHINE LEARNING ALGORITHMS

3.6.1 Decision Trees

A well-liked classification algorithm for simulating decisions and their outcomes is the decision tree. Decision trees are a useful tool in problem detection, as they can be used to categorize manufacturing batches according to a variety of features.

- **Algorithm:** A decision tree works by dividing a dataset into smaller groups according to the values of its features. This process results in a structure resembling a tree that represents different choices.
- **Example:** Using process parameters and sensor readings, a decision tree might identify batches as faulty or not.

Effectiveness: Although decision trees are easy to comprehend and straightforward, they can get overly complex and are prone to overfitting if not pruned appropriately [4].

3.6.2 Random Forests

Random forests utilize a technique to enhance the precision and resilience of classifications by merging several decision trees.

- **Algorithm:** It builds multiple decision trees using random subsets of features and combines their predictions to make a final decision.
- **Example:** Random forests can manage large datasets and identify complex patterns by aggregating the predictions from various decision trees.

Effectiveness: Random forests provide high accuracy and are less prone to overfitting compared to individual decision trees. They also manage missing values and noisy data well.

3.6.3 Support Vector Machines

Support vector machines (SVMs) are a powerful classification technique used to find the optimal boundary that separates different classes in the feature space.

- **Algorithm:** SVM constructs a hyperplane that maximizes the margin between different classes.
- **Example:** SVM can be used to classify batches based on sensor data by finding the best boundary that separates faulty and non-faulty samples [5].

Effectiveness: SVMs are effective in high-dimensional spaces and for complex datasets. However, they can be computationally intensive and require careful tuning of parameters.

3.6.4 Neural Networks

Deep learning models, in particular, which are utilized for intricate pattern recognition and fault detection tasks, are neural networks. They are made up of several layers of networked nodes, or neurons, which can recognize hierarchical characteristics in the data.

- **Algorithm:** They learn by adjusting weights through backpropagation to minimize the error between predicted and actual labels.
- **Example:** They are used by neural networks to minimize the error between the expected and real labels by learning via backpropagating weight adjustments.

Effectiveness: When it comes to finding intricate patterns and connections in big datasets, neural networks perform remarkably well. They do, however, need a large amount of training data and substantial processing power.

3.7 ANOMALY DETECTION TECHNIQUES

3.7.1 Principal Component Analysis

Principal component analysis (PCA) is a dimensionality reduction technique that can also be used for anomaly detection by identifying outliers in the principal component space.

- **Algorithm:** PCA projects the data onto a lower-dimensional space (principal components) and discovers outliers based on their distance from the mean in this reduced space.

- **Example:** PCA can spot abnormalities in sensor data by finding deviations from the principal components representing normal process behaviour.

Effectiveness: PCA is useful for reducing dimensionality and identifying anomalies in large datasets. However, it assumes linear relationships and may not capture complex anomalies [6].

3.7.2 Isolation Forest

An anomaly detection technique called isolation forest divides the data and chooses features at random to separate anomalies.

- **Algorithm:** By constructing an ensemble of isolation trees, the algorithm determines anomalies by measuring how simple it is to isolate them from the remaining data.
- **Example:** Isolation forest can detect faulty batches by identifying instances that are easily separated from the majority of normal data.

Effectiveness: Identifying abnormalities in high-dimensional datasets may be done effectively and efficiently with isolation forests. It works effectively with big datasets and does not require any presumptions on the distribution of the data.

3.8 HYBRID APPROACHES

Hybrid methods bring together different approaches to harness their advantages and overcome their drawbacks. For instance, merging machine learning algorithms with statistical techniques can improve the effectiveness of fault detection.

Effectiveness: By combining many techniques, hybrid approaches can offer more reliable and precise defect detection; nevertheless, they could need more intricate calibration and implementation.

To sum up, there are several methods and algorithms used in semiconductor production for fault detection, each having advantages and disadvantages. While machine learning algorithms enable more sophisticated pattern recognition and classification, statistical techniques like SPC offer fundamental monitoring capabilities. Techniques for detecting anomalies assist in locating odd patterns that may point to problems. Manufacturers can improve yield and process efficiency by combining these techniques to improve their ability to identify and fix defects.

3.9 DATA UNDERSTANDING AND PREPROCESSING

3.9.1 Detailed Analysis of the SECOM Dataset

The SECOM dataset contains a wealth of information from semiconductor manufacturing processes to aid in yield optimization and problem detection. It comprises a total of 1,567 entries, each representing a unique production batch or event. These samples are characterized by 591 attributes, encompassing variables, derived metrics related to the manufacturing process, and sensor readings. In addition to specific variables and calculated metrics, the features encompass continuous data from sensors that monitor numerous factors, such as temperature, pressure, and chemical concentrations. The dataset's target variable is binary with −1 indicating a pass and 1 indicating a failure, thereby denoting whether each batch successfully passed quality testing.

The analysis gains an extra layer of depth through the timestamp attached to each entry in the SECOM dataset revealing when the data was collected. This timestamp is invaluable for tracking events and identifying trends that could impact outcomes. In the dataset, initial observations often present a mix of feature distributions, with some showing significant fluctuations but others remaining stable. To enhance the accuracy of models, it is crucial to identify and address components with variance or prominent levels of noise.

Moreover, an analysis can uncover missing values in certain features that require careful handling to avoid distorting results. Additionally, the dataset may exhibit an imbalance between classes, with a notably smaller number of failed batches compared to successful ones. This imbalance can hinder the performance of algorithms, necessitating techniques such as weighted loss functions or resampling to achieve more precise predictions.

Taking everything into account, the SECOM dataset provides insights into the semiconductor manufacturing process and valuable information regarding factors affecting yield. By leveraging this data and employing appropriate feature selection and preprocessing techniques, researchers and engineers can develop robust models that enhance defect detection and optimize manufacturing processes.

3.9.2 Data Preprocessing

In the realm of preparing data for defect detection in semiconductor manufacturing, several key approaches are employed to ensure the dataset is ready for analysis. These strategies encompass filling in missing values, removing unnecessary features, and addressing zeros and *NaN* (not a number) values. Here is a detailed breakdown of these preprocessing activities:

a. **Handling Zeros and *NaN* Values:** Zeros and *NaN* values within datasets can signify incomplete or incorrect information. Given their potential impact on the quality of analysis, it is crucial to assess their presence and handle them appropriately. To address this, we begin by identifying the occurrences of zeros and *NaN*s across each column of the dataset. This process involves counting the quantity of these values for each feature to determine whether any columns contain inaccurate or missing data. Assessing the frequency of these values plays a role in informing the subsequent steps for data cleansing.
b. **Removing Constant Columns:** Some columns in datasets may have the same values across all rows. Because they lack diversity, these columns don't contribute to distinguishing between different outcomes or categories, making them unsuitable for analysis. To improve the effectiveness of the dataset, columns with identical entries are removed. This process involves checking if all the values in a column are identical. If a column meets this

criterion, it is excluded from the dataset. This step can enhance the efficiency of subsequent analyses or machine learning models by reducing complexity and eliminating irrelevant information.

c. **Imputing Missing Values:** Missing values in datasets pose a challenge that needs to be addressed for a thorough analysis. The first step in tackling missing values is identifying columns with a significant amount of missing data. Typically, columns with a higher percentage of missing values than a set threshold are discarded, as they may not provide reliable information.

To address the missing values in the dataset, imputation techniques are employed for the remaining columns. One commonly used approach involves utilizing the mean or median of the existing values in a column to replace any absent numbers. The mean provides a measure of tendency, while the median offers a metric that is less influenced by outliers. Ensuring that every feature has a value imputation helps maintain the integrity of the dataset. These preprocessing methods play a crucial role in preparing the data for analysis. By tackling zeros and *NaN*s, eliminating noninformative features, and filling in missing values, the dataset is refined and made more conducive to yielding trustworthy insights and predictions.

3.9.3 Feature Selection Techniques

A key component of data preparation is featuring selection, which finds the most pertinent features to enhance machine learning model performance. PCA, variance-based selection, and model-based selection are some of the methods used. Here is a thorough breakdown of every technique:

1. **Variance-Based Feature Selection:** The VarianceThreshold technique is employed to eliminate low-variance features. These features tend to remain consistent across samples, making their impact on the models' predictive capability minimal. By establishing a threshold, this approach removes attributes that do not meet the variance criteria. The goal is to retain only those elements that exhibit enough variation to be valuable. For instance, a feature would be removed from the dataset if its variance is below a specified threshold. This method reduces the dimensionality of the data and eliminates unnecessary noise [7].
2. **Model-Based Feature Selection:** This approach utilizes the effectiveness of machine learning models to assess the significance of each feature. The following models are utilized for this purpose.
   - **Support Vector Classifier (SVC):** This approach uses an L1 penalty to help pick out notable features. It relies on support vectors to spot the most relevant ones. The significance of the features is shown through the models' coefficients, and those that have coefficients get chosen.
   - **Logistic Regression:** This model looks at how features are related to the target variable. Features with coefficients are seen as more crucial.

By applying SelectFromModel along with logistic regression, it selects features that carry great importance.
- **Decision Tree Classifier:** Decision trees evaluate the significance of features based on how effectively they divide the data. Features that result in splits are considered more significant. The ExtraTreesClassifier, which is a type of decision tree, combines outcomes from multiple trees to determine feature selection.

3. **Principal Component Analysis:** PCA is a technique used to reduce the dimensionality of data by transforming it into a new coordinate system. In this system, the largest deviations in the data projections align with the initial principal components. PCA achieves this by minimizing features while preserving a massive portion of the original dataset variability. The process involves projecting the data onto components that are derived from linear combinations of the original features. The result is a set of components or uncorrelated features that capture most of the data variance.

In summary, employing selection methods like selection based on variance, model-based selection, and PCA are essential for enhancing the effectiveness and efficiency of learning models. These approaches can boost the performance of models, highlight key aspects, and simplify data complexity [8].

3.9.4 Oversampling for Imbalanced Datasets

When it comes to classification tasks, datasets often suffer from an imbalance, where one class outweighs the other significantly. This disparity can lead the model to lean towards the majority class, resulting in biased performance and poor accuracy predictions for the minority class. To tackle this issue, oversampling techniques like the synthetic minority oversampling technique (SMOTE) are employed to equalize the class distribution. SMOTE allows for the creation of samples specifically for the underrepresented class within a dataset. Instead of simply duplicating existing minority class instances, SMOTE generates new synthetic samples through interpolation between existing data points. This process involves selecting points and generating additional samples along the lines connecting these points to their neighbouring instances. By doing so, SMOTE effectively boosts the representation of the minority class, leading to a more balanced dataset overall.

To determine how classes are distributed, the approach begins by examining the initial dataset. In cases of imbalance, the minority class may be underrepresented, leading to a flawed model. Subsequently, SMOTE is applied to the dataset to generate additional instances of the minority class. This adjustment helps create a foundation for training by balancing the number of samples across classes. The modified dataset, with an equal number of samples for each class after SMOTE, is utilized for further analysis or model training. This balance boosts the model's performance and reduces bias by enhancing its ability to generalize across all classes. With the support of the newly generated synthetic samples, the model gains insights into the minority class, improving prediction accuracy and overall robustness.

SMOTE is an approach to address imbalanced classes in datasets. By generating examples, it ensures that the model is trained on a dataset, leading to improved performance for both majority and minority classes. This technique is crucial for developing unbiased and precise models, especially in scenarios where outcomes can be significantly influenced by class imbalance [9].

3.10 MODEL EVALUATION AND VISUALIZATION

3.10.1 Training and Testing the Model

The model evaluation process starts with training a classification model on a dataset. In this instance, we use a random forest classifier, which combines multiple decision trees and merges their results to boost accuracy and reliability. We split the dataset into training and testing parts using a set test size, making sure we can assess the model on data it has not seen before.

Once we have trained the model, we make predictions on the test set. We then gauge the model's performance using accuracy, which shows how many predictions were correct out of all the predictions made. The classifier function gives back the trained model, predictions, actual labels, and accuracy score.

3.10.2 Confusion Matrix Calculation

A confusion matrix helps us check how well a classification model works. It breaks down the model's guesses in detail, showing true positives, true negatives, false positives, and false negatives. To figure out these numbers, the confusion_matrix function compares the labels the model predicted to the real labels.

Here is how the confusion matrix is set up:

True Positives (TP): Instances where the model correctly predicted the positive class.

True Negatives (TN): Instances where the model correctly predicted the negative class.

False Positives (FP): Instances where the model incorrectly predicted the positive class.

False Negatives (FN): Instances where the model incorrectly predicted the negative class.

These values are used to construct a matrix that visually represents the performance of the model.

3.10.3 Visualization of the Confusion Matrix

To understand the confusion matrix better, we show it as a heat map. The plot_confusion_matrix function creates this visual. The heat map gives us a clear picture of how well the model works. Each box in the matrix shows how many times the model predicted a class compared to the actual class.

We can display the matrix with or without normalization. Normalization changes the values to percentages, which helps when comparing different datasets or models. The function uses colours to make things clearer, and it puts the exact numbers in each box for more detail.

To wrap up, evaluating a model involves three steps: training it, getting its predictions, and checking how well it used a confusion matrix. This matrix sheds light on what the model does well and where it falls short. Seeing it makes it easier to grasp the results. This method gives us a full picture of how good the classification model is and ensures a comprehensive evaluation of the classification model's performance.

3.10.4 Feature Selection and Classification

The first step in the process is to get the dataset ready. This involves cleaning up any missing data. Next, we use the SVC to choose features. It finds and keeps the most important traits for sorting. After that, we balance the picked features using SMOTE. We do this to fix any class imbalance in the dataset. This step ensures the classifier has a balanced dataset to work with.

The random forest classifier, trained on balanced data, classifies the samples. To evaluate the model's performance, we calculate its accuracy score and analyse the data using a confusion matrix. This matrix, which we visualize with normalization, gives us insight into the model's strong and weak points by showing true positives, true negatives, false positives, and false negatives. This thorough assessment helps us understand how well the model works and spot areas we can make better. Figure 3.1 illustrates the normalized confusion matrix for SVC feature selection with random forest classification.

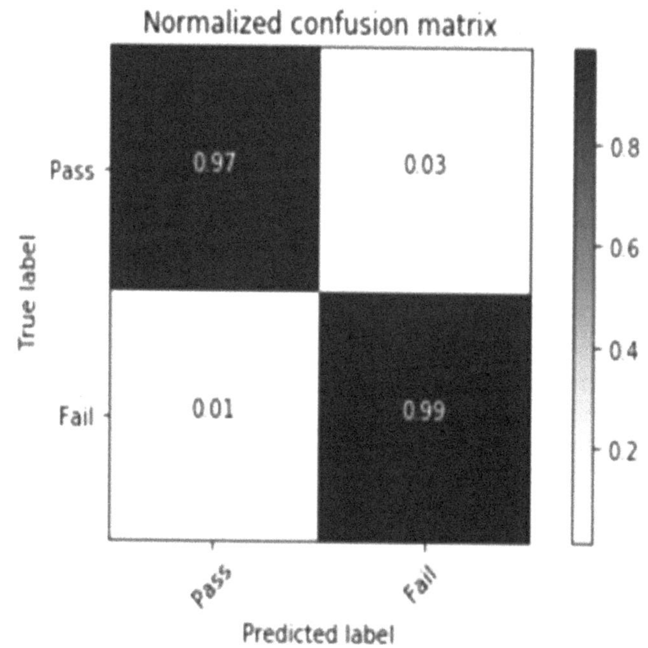

FIGURE 3.1 Normalized confusion matrix for SVC feature selection with random forest classification.

In the same way, once we get the dataset ready and clean it up, we use a logistic classifier to pick out the best features. This classifier checks which features are most likely to predict the outcome, which helps us narrow down the dataset to the most useful variables. After this, we balance the dataset again using SMOTE. This makes sure the classifier learns from an even spread of classes.

The balanced dataset serves as input to train a random forest classifier. We measure the classifier's performance using accuracy and create a normalized confusion matrix to assess the model's classification abilities. The confusion matrix, which we visualize with normalization, shows how well the model differentiates between classes, including true positives, true negatives, false positives, and false negatives. This evaluation gives us a clear picture of how the model performs and points out areas we can improve and optimize. The normalized confusion matrix for logistic feature selection with random forest classification is shown in Figure 3.2.

A decision tree model helps pick out key features after the dataset is ready and cleaned up. This method checks how important different traits are, making it easier to spot the ones that matter the most for predicting outcomes. By focusing on the most influential factors, the classification process becomes more productive and effective, thanks to the chosen features.

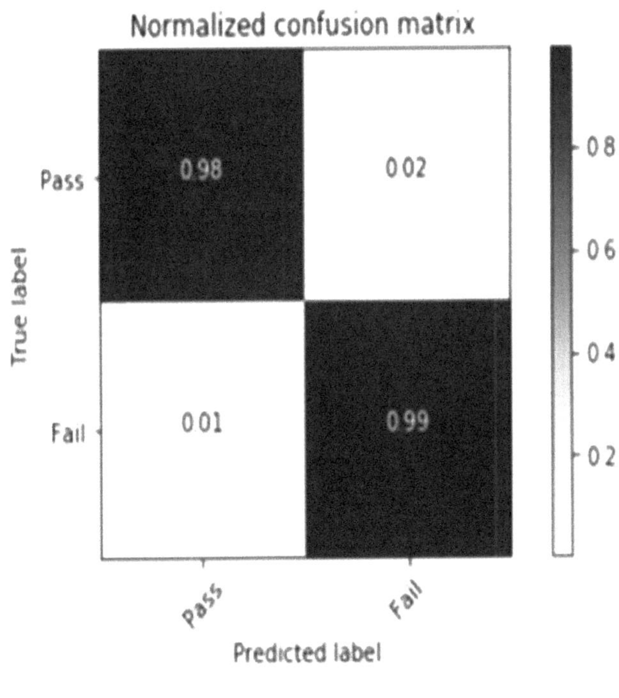

FIGURE 3.2 Normalized confusion matrix for logistic feature selection with random forest classification.

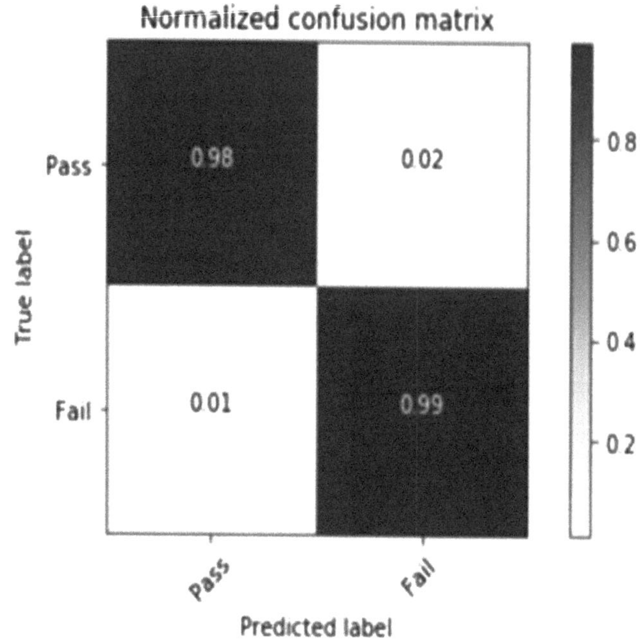

FIGURE 3.3 Normalized confusion matrix for decision tree feature selection with random forest classification.

After picking the features, we tackle any uneven class spread by evening out the dataset using SMOTE. This balancing step makes sure the random forest classifier learns from an equal number of each class, which is key to building a trustworthy and fair model.

After this, we use this balanced dataset to train the random forest classifier. We use a normalized confusion matrix to visualize how accurate the model is, which helps us evaluate its performance. This matrix gives us a thorough analysis of how well the classifier does, showing us the number of false positives, false negatives, true positives, and true negatives. By normalizing the confusion matrix, we can compare the model's performance across different classes more easily, which helps us identify where the model shines and where it falls short. Figure 3.3 highlights the normalized confusion matrix for decision tree feature selection with random forest classification.

3.11 RESULTS AND DISCUSSION

Examining the different feature selection and classification methods using the SECOM dataset gives us fresh insights into their effectiveness for spotting defects in semiconductor manufacturing. We applied SVC, logistic regression, and decision trees to select features and then used random forest models to classify them. This approach helped us gain a deep understanding of how various techniques affect performance.

The random forest model got better at sorting things out when we trimmed down the data to just the key bits, using SVC to pick what mattered most. Looking at the results in an obvious way showed that the model was good at telling pass from fail, which means the SVC way of choosing what to look at made the model work better. By spotting the most important things to consider, this method produced predictions you could count on more.

Similarly, using logistic regression to pick features worked well. The random forest classifier did an excellent job after training on the chosen features. The confusion matrix showed some wrong guesses, both false positives and negatives, but overall, the classification was pretty accurate. This tells us that while logistic regression was good, we might be able to make it work even better to catch subtler patterns in the data.

The decision tree method of picking features gave us another helpful view. It ranked features based on how important they were, which helped the random forest classifier do well. The confusion matrix showed that the decision tree was good at spotting key aspects. It had a nice balance between correct positive and negative classifications.

Looking at the outcomes of these different methods sheds light on the upsides of each feature selection approach. SVC-based feature selection did a fantastic job of refining the dataset, while decision trees and logistic regression brought their own benefits to the table. Each technique revealed areas that could use some tweaking to boost performance but also showed its strengths in making the classification more accurate. Down the road, researchers should focus on exploring new ways to select features and fine-tune algorithms to achieve even better accuracy.

3.12 CONCLUSION

The study of feature selection methods paired with random forest classification for the SECOM dataset showed big improvements in spotting defects. SVC, logistic regression, and decision trees each played a key role in boosting the model's effectiveness. SVC provided a refined dataset with high classification accuracy, while logistic regression and decision trees also produced satisfactory results, each with its own strengths and weaknesses. Concisely, the research highlights how important it is to pick the right feature selection method to get the best model performance. Comparing different approaches gives insights that show how well feature selection works to improve fault detection systems. Looking ahead, future work should focus on making current methods better and exploring new ones to achieve even higher accuracy and reliability in semiconductor manufacturing.

REFERENCES

1. J. Shim, S. Cho, E. Kum, and S. Jeong, "Adaptive fault detection framework for recipe transition in semiconductor manufacturing," *Computers & Industrial Engineering*, vol. 161, 2021.
2. S. Mitra, and C. Koley, "Real-time robust bearing fault detection using scattergram-driven hybrid, CNN-SVM," *Electrical Engineering*, pp. 1–11, 2023.

3. T. Schlosser, M. Friedrich, F. Beuth, and D. Kowerko, "Improving automated visual fault inspection for semiconductor manufacturing using a hybrid multistage system of deep neural networks," *Journal of Intelligent Manufacturing*, vol. 33, no. 4, pp. 1099–1123, 2022.
4. M. Glučina, N. Anđelić, I. Lorencin, and Z. Car, "Detection and classification of printed circuit boards using YOLO algorithm," *Electronics*, vol. 12, no. 3, p. 667, 2023.
5. X. Yu, L. Han-Xiong, and H. Yang, "Collaborative learning classification model for PCBs defect detection against image and label uncertainty," *IEEE Transactions on Instrumentation and Measurement*, vol. 72, pp. 1–8, 2023.
6. T.J. Kutre, S.N. Patil, S. Kore, and V.M. Aparanji, "Advanced architecture of analog to digital converter derived from half flash ADC," in *Distributed Computing and Optimization Techniques: Select Proceedings of ICDCOT 2021*, pp. 141–151, 2022.
7. I.H. Ozcan, O.C. Devecioglu, T. Ince, L. Eren, and M. Askar, "Enhanced bearing fault detection using multichannel multilevel 1D CNN classifier," *Electrical Engineering*, vol. 104, no. 2, pp. 435–447, 2022.
8. M. Sen, and C. Gayatri, "FReNG: Federated optimization by using regularized natural gradient descent," in *Accepted at International Conference on Machine Learning and Applications (ICMLA)*, December 15–17, 2023.
9. K. Xia, Z. Lv, K. Liu, Z. Lu, C. Zhou, H. Zhu, et al., "Global contextual attention augmented YOLO with ConvMixer prediction heads for PCB surface defect detection," *Scientific Reports*, vol. 13, no. 1, p. 9805, 2023.

4 Predictive Modelling for Yield Enhancement

Arpita Pattnaik, Neeraj Gupta, Rashmi Gupta, Arun Kumar Singh and Rajkumar Sarma

4.1 INTRODUCTION

The percentage of defect-free chips produced is known as yield, and it is optimized to the highest degree in the hectic world of semiconductor manufacturing. The semiconductor industry is constantly working to streamline production procedures to minimize costs and time to market and satisfy the growing demand for high-quality chips. Achieving the ideal yield, however, is a difficult task affected by a wide range of variables, such as equipment performance, material qualities, and process variances.

Historically, statistical process control (SPC), advanced process control (APC), and other statistical techniques have been utilized in yield improvement initiatives to track and modify industrial processes. These methods have proven helpful in locating and reducing process variances, but they frequently miss the complex connections that exist between process variables and yield results.

Deep learning – a subfield of artificial intelligence (AI) – which takes its cues from the structure and operations of the human brain, has grown in prominence in recent years and completely changed a number of industries, including semiconductor production. Deep learning models present a possible path for improving predicted yield in semiconductor production due to their ability to automatically extract complex relationships and patterns from massive amounts of data. As algorithms for training and inference become more efficient, deep learning is finding application in a variety of disciplines, including microcontrollers and data centres. Deep learning methods like generative models and evolutionary algorithms are being used in semiconductor manufacturing for many processes, including process optimization, automated optical inspection, and photomask manufacture. Because of the volume of data, calculating the yield across all process steps in detail is an extremely complicated task that calls for either an arbitrary number of resources for a brute force study or an effective heuristic approach. Currently, historical data can be used to make inferences about the frequency of errors occurring and the links between errors; unfortunately, not all errors, process configurations, and parameter combinations are documented. As a result, it is impossible to identify errors without gaps [1]. Realizing the full potential of AI semiconductor technologies in this field requires overcoming obstacles such as comprehending the shift from training to inference and optimizing computations.

DOI: 10.1201/9781003508304-4

This chapter examines the methods using deep learning to forecast and maximize yield in semiconductor production. In one study, principal component analysis is used to estimate yield by focusing on test parameters for non-normal distributions [2]. We explore the essential elements of this methodology, encompassing data gathering, preprocessing, model building, training, implementation, and ongoing enhancement. Semiconductor makers can get important insights into their manufacturing processes, pinpoint variables influencing yield, and put focused strategies into place to raise overall yield and productivity by utilizing deep learning.

4.2 NEED OF DEEP LEARNING IN YIELD ENHANCEMENT

Conventional methods for improving yield in semiconductor production have been established over many years and combine engineering principles, process control strategies, and statistical methodologies. These methods have been essential in preserving process stability, enhancing overall yield, and optimizing parameters. They do, however, have several drawbacks and restrictions that may reduce their efficacy.

4.2.1 Overview of Statistical Process Control

- Statistical process control is a methodology that uses statistical analysis of process data to monitor and control manufacturing processes. A disadvantage of SPC is that, although it is good at identifying fluctuations in the process and preserving stability, it might not be able to capture intricate, nonlinear correlations between the process parameters and yield results.
- Impact: This restriction may make it more difficult to spot minute dependencies and interactions between variables, which could result in less-than-ideal process optimization and yield improvement.

4.2.2 The Design of Experiments

- Design of experiments (DOE) is a systematic methodology that aims to enhance product quality and performance by examining and refining process factors. Although DOE is helpful for process optimization, it is frequently reactive in character, concentrating on finding and fixing problems after they arise.
- Impact: The capacity to proactively minimize yield loss may be limited by this reactive strategy, which may lead to an increase in downtime, scrap, and rework.

4.2.3 Advanced Process Control

- The goal of advanced process control is to maintain desired performance by adjusting process parameters through the integration of sensors, control algorithms, and process models. APC systems might not be able to manage the enormous amounts of data produced by contemporary semiconductor fabrication techniques.

- Impact: This restriction may make it more difficult to evaluate intricate datasets with a variety of information sources, which could lead to the loss of important information that could be used to increase yield.

4.2.4 Failure Analysis

- Failure analysis, also known as root cause analysis, is the methodical examination of flaws or failures, with the goal of locating the underlying reasons and averting recurrence. The drawback is that although failure analysis is crucial to comprehending flaws, it is frequently a labour-and resource-intensive procedure.
- Impact: Scalability and agility may be restricted by manual analysis, which would impede prompt fault prevention and prolong remedial operations.

4.2.5 Overview of Process Optimization and Continuous Improvement

- To improve performance and yield, manufacturing processes must be consistently improved through process optimization. A disadvantage of traditional process optimization methods is that they might not be able to predict future yield trends.
- Impact: Manufacturers can find it difficult to reach production goals and maintain steady yield rates in the absence of predictive skills.

While conventional methods are still useful tools in the semiconductor manufacturing industry, combining them by using state-of-the-art technology, such as machine learning and advanced analytics, can help us get around some of their drawbacks and open up new possibilities for process and yield optimization.

4.3 ADVANTAGES OF DEEP LEARNING FOR YIELD IMPROVEMENT

Here are some benefits of applying deep learning to yield improvement and predictive modelling in semiconductor manufacturing.

4.3.1 Ability to Capture Complex Interactions

- The multilayered structures of deep learning models enable them to efficiently depict the intricate and nonlinear relationships that exist between yield outcomes and process parameters.
- They can uncover intricate connections and patterns in the data that may be difficult for conventional statistical techniques to identify.

4.3.2 Feature Learning and Representation

- Manual feature engineering is no longer necessary because deep learning algorithms can automatically extract relevant features from raw data.
- The capacity to acquire significant data representation can result in prediction models that are more reliable and accurate.

4.3.3 Managing Big Data

- Deep learning models are highly effective at tackling the big data difficulties present in semiconductor production because of their prowess in processing and interpreting massive datasets.

4.3.4 Flexibility

- Deep learning frameworks offer a great degree of flexibility and adaptability to diverse data kinds and problem areas.
- Because of its capacity to support a vast amount of data sources and formats, semiconductor makers are able to incorporate data into their predictive models from many sources.

4.3.5 Increased Predictive Accuracy

- In a range of applications, deep learning frameworks, especially their sophisticated structures, such as convolutional neural networks (CNNs) and recurrent neural networks (RNNs) have demonstrated better predictive accuracy than traditional statistical models.
- This increased accuracy can result in more accurate yield projections and better decision-making in the semiconductor manufacturing industry.

4.3.6 Real-Time Prediction

- By continuously analysing incoming data from manufacturing processes, deep learning frameworks can be used in real time to generate prompt yield-quality predictions.
- Proactive decision-making and prompt interventions make it possible to optimize process parameters and avoid yield loss.

4.3.7 Identification of Anomalies and Abnormalities

- Deep learning models can be trained to find process deviations, material flaws, and equipment faults, which are examples of anomalies and abnormalities found in semiconductor manufacturing processes.
- Production disruptions can be minimized and yield loss avoided with the early detection of these problems.

4.3.8 Efficiency and Scalability

- Deep learning models are scalable and deployable on a large scale across manufacturing facilities once trained.
- This enables reliable yield prediction across many production lines and locations. Because of its scalability, semiconductor makers can use predictive modelling to increase yield throughout their whole business.

Semiconductor makers can get important insights into their manufacturing processes, pinpoint variables influencing yield, and put focused strategies into place to raise overall yield and productivity by utilizing deep learning. The current methods for using metrology data to classify yields emphasize counteracting imbalanced classes and imputing missing data [3]. The majority of real-world problems require the use of advanced analytics [4] so that thoughtful and intelligent solutions can be developed to meet current needs. For this reason, analytical AI that makes use of machine learning and deep learning approaches can be crucial to the development of AI-powered computing and systems.

4.4 FUNDAMENTALS OF DEEP LEARNING

Gaining an understanding of the foundations of deep learning is crucial to maximizing its potential to increase yield in semiconductor production. Let's examine each of the main ideas in more detail.

4.4.1 Neural Networks

- The fundamental architecture of a neural network is a linked network of nodes arranged in layers. The output layer generates the model's predictions, the hidden layers carry out calculations, and the input layer receives raw data. Activation functions provide nonlinearity to the network and enable it to simulate intricate correlations in the data; nonlinear functions are applied to the output of every neuron. Some examples of common activation functions include Tanh, sigmoid, and ReLU (rectified linear unit). During the training phase, weights and biases are parameters related to the connections between neurons that are changed to reduce the discrepancy between the expected and observed results.
- Gradient boosting machines (GBMs) have become potent tools for semiconductor manufacturing, process condition optimization, and material selection. In terms of probabilistic estimates, uncertainty estimates, and predictive performance, these machine learning models frequently beat deep neural networks. GBMs are particularly competent at things like process condition optimization for different stages of semiconductor manufacture, such as lithography, etching, and deposition. Process engineers looking to increase yield and decrease variability can benefit greatly from their capacity to offer precise forecasts and trustworthy estimates of uncertainty.

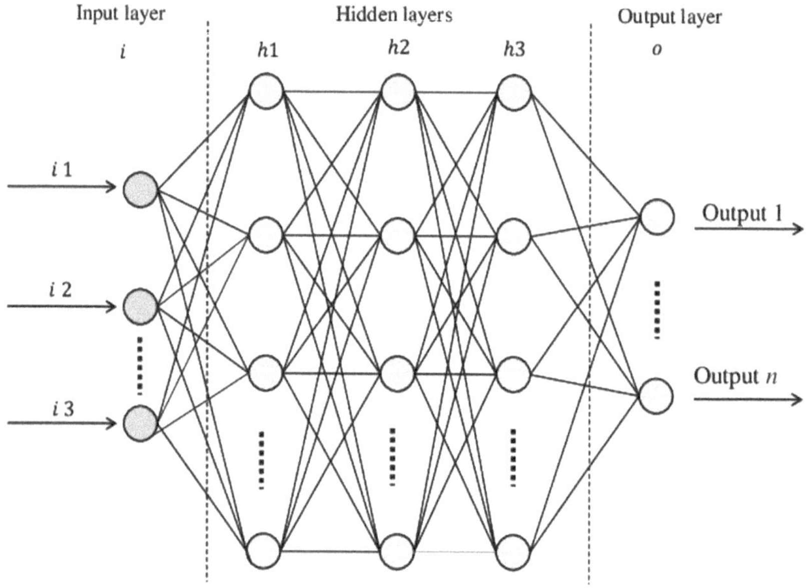

FIGURE 4.1 FNN architecture.

Source: Copyright © www.researchgate.net.

4.4.2 Deep Learning Architectures

- **Feedforward Neural Networks:** A feedforward neural network (FNN) is the most basic neural network type, where data only moves from input to output in a single direction, devoid of loops or cycles. They work well for classification and regression applications. Figure 4.1 shows the FNN architecture.
- **Convolutional Neural Networks:** CNNs are designed specifically to manage information that resembles a grid, like photographs. Convolutional, pooling, and fully connected layers make up their structure, which allows them to record patterns and hierarchies in space. Figure 4.2 depicts the CNN architecture.
- **Recurrent Neural Networks:** RNNs are made to manage temporally dependent sequential data. They are suitable for applications like time-series analysis and natural language processing prediction because they have feedback loops that enable information to endure across time. Figure 4.3 shows the RNN architecture.
- **Long Short-Term Memory (LSTM) Networks:** The LSTM network is an RNN design that uses memory cells to solve the vanishing gradient issue. Long short-term dependencies can be learned via LSTMs, which are popular for sequential tasks.
- **Transformer Models:** Transformer models are a new development in deep learning architecture mostly applied to challenges involving natural

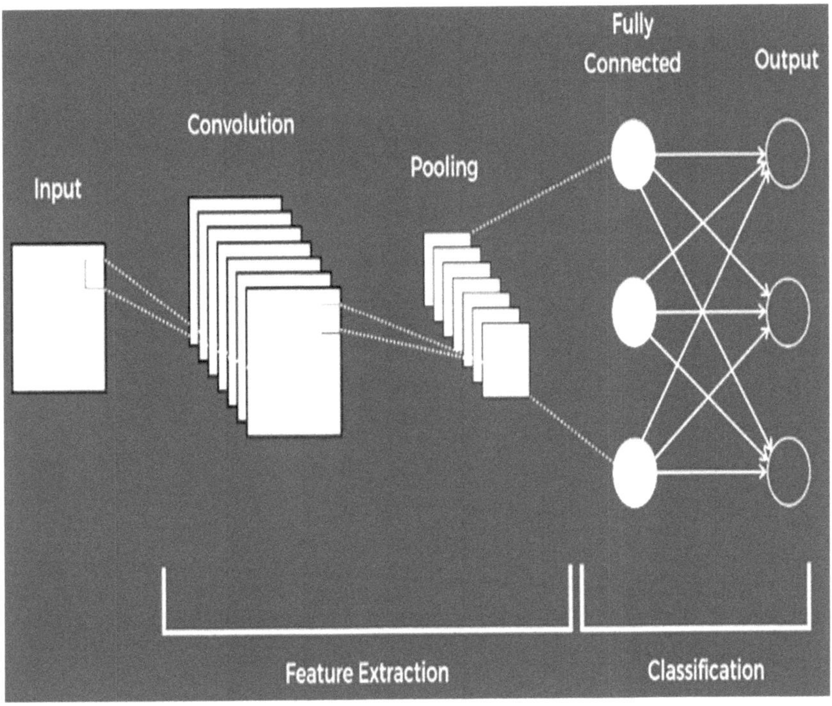

FIGURE 4.2 CNN architecture.

Source: Copyright © medium.com.

language processing. Transformers efficiently capture global relationships in sequential data by using self-attention methods.

4.4.3 Instruction Models for Deep Learning

- The process of sending input data through a network to calculate predictions is called forward propagation. After performing computations with its own weights and biases, each layer transfers the results to the subsequent layer.
- Backpropagation is a method of changing the weights and biases in the model in response to variations between the expected and actual results. It involves calculating the gradients of the loss function with respect to each parameter and altering them using optimization methods such as Adam or stochastic gradient descent (SGD).
- Loss functions are objective functions that express how much the actual and projected outcomes differ from one another. Cross-entropy loss is a popular loss function for classification applications, whereas mean squared error (MSE) is employed in regression tasks.

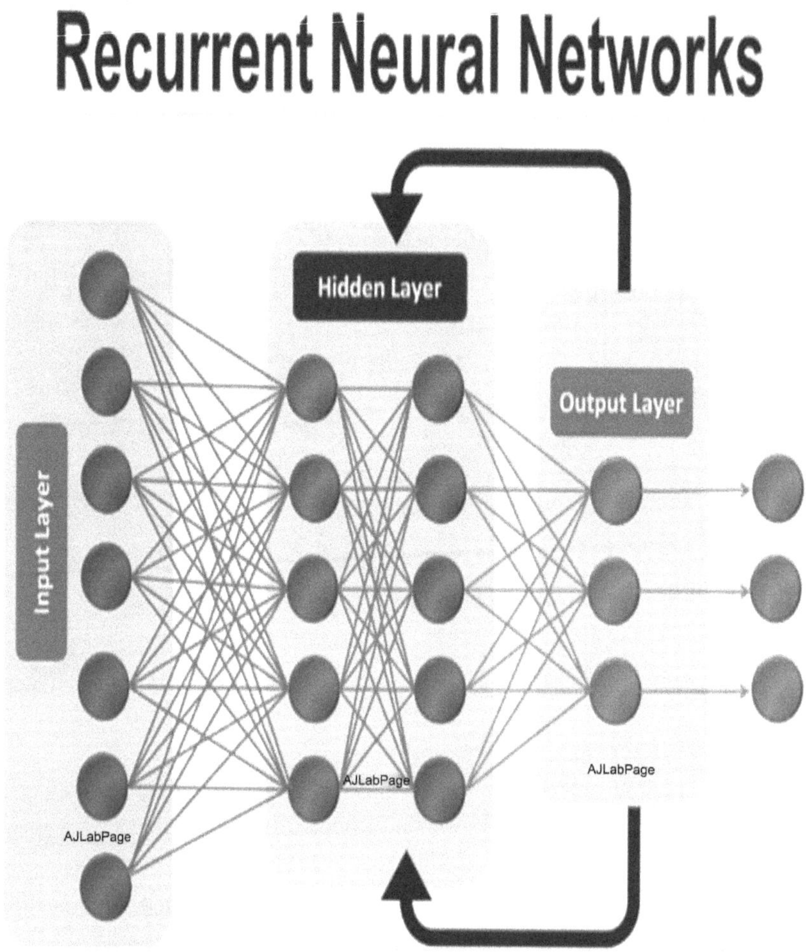

FIGURE 4.3 RNN architecture.

Source: Copyright © www.researchgate.net.

4.4.4 Adjusting Hyperparameters and Models

- **Learning Rate:** It manages the step size when updating parameters. Selecting a suitable rate of learning is essential for effective training and convergence.
- **Batch Size:** The quantity of samples handled in each training iteration is determined by the batch size. It has an impact on both memory needs and the stability of the training process.

Quantity of Neurons and Layers: The neural network's architecture, which includes the quantity of layers and every neuron's layer, affects the network's ability to recognize intricate relationships and patterns.

- **Methods of Regularization:** Techniques like batch normalization, dropout, and L1/L2 regularization aid in preventing overfitting and enhancing the framework's capacity for generalization.

4.4.5 Model Evaluation and Validation

Usually, data is divided into several subsets for testing, validation, and training. The training set is used to train the model, performance is monitored, the validation set's hyperparameters are changed, and the final model's effectiveness is evaluated on the testing set. Deep learning models' performance on classification and regression tasks is measured using several metrics, including accuracy, precision, recall, F1 score, and mean absolute error (MAE). A wide range of knowledge categories, including procedural, meta-, structural, descriptive, and heuristic knowledge, can be applied in different application areas [5].

4.4.6 Model Utilization and Interpretation

Deep learning models can be implemented on multiple platforms, such as embedded systems, edge devices, and cloud-based services. Making predictions on new, unseen data using a trained model is known as inference. It is the process of using a trained model to process input data and produce predictions as output. A knowledge-based conceptual model can be developed using a variety of knowledge representation techniques, such as logical, semantic, network, frame, and production rules [6].

4.5 DATA COLLECTION AND PREPROCESSING

A thorough grasp of the procedures involved is crucial when exploring data collection and preprocessing in the context of deep learning–based predictive modelling for semiconductor yield enhancement. Here's a thorough explanation.

- **Data Collection:** Sensor data such as temperature, pressure, flow rates, chemical concentrations, and other characteristics are monitored by a variety of sensors during the semiconductor manufacturing process. Sensor types include flowmeters, thermocouples, pressure sensors, and spectroscopy equipment.
- **Data Granularity:** Depending on the process, sensor data is normally collected at regular intervals, which can be anywhere from milliseconds to minutes.
- **Sources of Data:** Manufacturing equipment provides this data, which is then gathered and kept in databases, historical repositories, and data warehouses.
- **Equipment Logs:** The machinery utilized in the fabrication of semiconductors produces logs that record data regarding performance, maintenance, and operations. Events, alerts, recipes, and equipment status logs are among the several types of logs.

- **Data Forms:** Logs might be kept in XML, CSV, or other proprietary formats that are unique to certain equipment manufacturers.
- **Data Cleaning:** To extract pertinent features from logs, which frequently contain noise and irrelevant information, preprocessing is necessary.
- **Inspection and Metrology Findings:** The quality and integrity of semiconductor wafers and chips are evaluated using metrology instruments and inspection systems. Measurements of critical dimensions, defect counts, defect maps, and wafer maps are examples of data types.
- **Data Integration:** For a thorough analysis, inspection and metrology data must be combined with other process data.
- **Alignment:** For wafer maps and defect maps to match process data, alignment and registration may be necessary.
- **Past Information:** Predictive models can be effectively trained using historical data from prior manufacturing runs.
- **Data Quality:** Careful preprocessing is necessary since historical data may have missing values, inconsistent information, or out-of-date information.
- **Preprocessing the Data:** It involves cleaning the data and managing missing values. Depending on the type of missing values, methods like imputation, deletion, or interpolation are used.
- **Outlier Removal and Detection:** This step involves finding and eliminating outliers that could distort the results of an analytic or modelling procedure.
- **Noise Reduction:** Methods like low-pass filters or moving averages remove noise from sensor data.
- **Scaling and Normalization:** To guarantee consistency and enhance convergence during model training, numerical characteristics are scaled and normalized to a standard range (such as [0, 1]).
- **Standardization:** To account for differences in scale and variance, the characteristics are adjusted to have a standard deviation of 1 and mean of 0.
- **Dimensionality Reduction:** Methods of reducing dimensions, such as feature selection or principal component analysis, are employed in characteristic engineering to focus on the most pertinent characteristics by reducing the dimensionality of the dataset.
- **Features of Time Series:** From time-series data, temporal elements are extracted, including rolling averages, lag values, and trend markers.
- **Features Particular to a Domain:** Features are engineered with a focus on semiconductor production processes and domain knowledge.
- **Integration of Data:** To enable insightful analysis and accurate modelling, it is essential to ensure that various data sources (such as sensor data, equipment logs, and metrology findings) are consistent and aligned with one another.
- **Joining and Aggregation:** By utilizing relevant keys or IDs to merge data from several sources, data must be combined at the desired granularity (e.g., wafer level, lot level).

- **Data Splitting:** To effectively assess the performance of the model, the dataset must be divided into distinct subsets for testing, validation, and training.
- **Time-Based Splitting:** To prevent data leaks and evaluate model generalization, temporal dependencies must be taken into account by dividing data into chronological chunks.
- **Data Augmentation:** To increase the resilience and generalization of the model, more training data can be generated by rotating, flipping, or adding noise.

Building accurate and trustworthy predictive models for yield enhancement in semiconductor production requires effective data collection and preprocessing. Practitioners can ensure that their models are trained on high-quality data and are able to capture pertinent patterns and correlations in the manufacturing process by carefully selecting and preparing the data.

4.6 MODEL DEVELOPMENT AND TRAINING

Using deep learning for predictive modelling in semiconductor production requires a thorough understanding of the model building and training stages. Now let's examine these procedures in more detail.

- **Model Development:** Clearly identify the goals of the predictive model as well as the problem statement. In semiconductor manufacturing, yield augmentation may entail process anomaly detection, production parameter optimization, or defect occurrence prediction.
- **Data Preparation:** Compile and prepare the information needed to train the model. This entails gathering information from a variety of sources, including sensors, equipment records, and inspection findings and then performing feature engineering, data standardization, and data cleansing.
- **Model Selection:** Based on the type of data and the issue at hand, select a suitable deep learning architecture. Transformer frameworks, RNNs, feed-forward neural networks, and CNNs are examples of common architectures.
- **Design of Architecture:** Describe the neural network's design, taking into account the number of layers, kinds of layers (such as recurrent and convolutional), and mechanisms of activation that are utilized in each layer.
- **Training of the Model:** Divide the dataset into distinct portions for testing, validation, and training. The validation set is used to track the performance and adjustment of hyperparameters, the testing set is used to assess the performance of the finished framework, and the training set is used to update the model's parameters.
- **Selection of Loss Function:** Depending on the requirements of the predictive model task, select a suitable loss function. MSE and MAE are popular loss functions for regression tasks; cross-entropy loss is frequently employed for classification problems.

- **Hyperparameter Tuning:** To maximize model performance, adjust hyperparameters, including learning rate, batch size, number of epochs, and regularization parameters. To determine the ideal set of hyperparameters, employ strategies like automated hyperparameter optimization methods, random search, or grid search.
- **Instructional Procedure:** To calculate predictions for the training set, use forward propagation. To calculate the loss function's gradients in relation to the model parameters, use backpropagation. Then use algorithms for optimization like Adam, RMSprop, or SGD to update the parameters. Metrics like training loss and validation loss can be used to track training progress and identify instances of overfitting or underfitting.
- **Regularization:** To avoid overfitting and enhance the framework's capacity for generalization, use regularization strategies such as batch normalization, dropout, and L1/L2 regularization.
- **Early Stopping:** As the model performs on the validation set, monitor its process throughout the training, and stop the model when it deteriorates, indicating overfitting. To avoid overfitting, save the model parameters that correspond to the lowest validation loss.
- **Model Evaluation:** Using suitable evaluation metrics, such as accuracy, precision, recall, F1 score, or MAE, assess the final trained model on the testing set. To evaluate the capacity of the model to generalize, compare its performance on the training and validation sets with that of the testing set.
- **Iterative Refinement:** Based on insights from the model evaluation and input from subject matter experts, iterate on the model construction and training process.

To increase the robustness and performance of your model, experiment with various preprocessing methods, hyperparameters, and architectures. Practitioners can significantly increase production efficiency and product quality by carefully following these steps to develop and train deep learning models that accurately capture the intricate relationships found in semiconductor manufacturing data and predict yield outcomes.

4.7 DEPLOYMENT OF PREDICTIVE MODELS

To use predictive models in semiconductor manufacturing, development settings must give way to production settings while maintaining real-time performance, scalability, and dependability. This is a thorough rundown of the deployment procedure.

4.7.1 Infrastructure Configuration

- **Hardware Needs:** Establish the gear, such as servers, graphics processing units, or specialized hardware accelerators for deep learning inference, required to support the deployment of the model.

- **The Software Stack:** Install the necessary software stack, which should include any extra monitoring and management tools, data preprocessing and model serving libraries (such as TensorFlow Serving, TorchServe), and deep learning frameworks (such as TensorFlow, PyTorch).

4.7.2 Model Packaging

- **Model Serialization:** Serialize the deep learning framework that has been trained into a format that can be used for deployment, like the SavedModel format from TensorFlow or the TorchScript format from PyTorch.
- **Dependencies Management:** Ensure the model is deployed with all necessary dependencies, such as Python packages, deep learning frameworks, and custom libraries.

4.7.3 Model Deployment

- Using a model serving framework or platform, deploy the serialized model to the production environment. This can entail putting the model into use as a Docker container, RESTful API, or cloud provider inference service.
- **Scalability:** Ensure the deployment infrastructure has the capacity to grow to meet the changing demands for inference requests and handle a range of workloads.

4.7.4 Combining Production Systems with Integration

- Set up systems for the ingestion of real-time data, such as equipment logs, sensor data streams, and inspection findings. Preprocess the data before supplying it to the deployed model.
- **Combining Manufacturing Systems:** To facilitate smooth data flow and decision-making, integrate the predictive model with the current production processes and systems, such as production execution systems, process control systems, and quality management systems.

4.7.5 Observation and Record-Keeping

- **Monitoring Model Performance:** Use monitoring tools to keep an eye on the deployed model's performance in real time, tracking variables like error rates, latency, throughput, and resource usage.
- **Logging and Auditing:** To aid in debugging, auditing, and troubleshooting, log inference requests and replies in addition to metadata like timestamps, input data, and model predictions.

4.7.6 Safety and Adherence

- **Data Security:** To safeguard private information sent to and from the deployed model, use communication protocols for security purposes, accessing the controls, and data encryption.

- **Compliance:** Verify that the deployment complies with legal requirements as well as industry standards pertaining to data security, privacy, and quality control, including ISO 27001, GDPR, HIPAA, and SEMI standards.

4.7.7 Constant Inspection and Upkeep

- **Optimizing Performance:** To increase accuracy and efficiency over time, keep an eye on the deployed framework's execution, and adjust the hyperparameters, change the architecture of the model, or retrain the frameworks using new data.
- **Fault Tolerance:** Use techniques like redundancy, failover, and automatic scaling to preserve system resilience and manage failures gracefully.
- **Versioning and Rollback:** Keep track of deployed model versions, and put methods in place to roll back to earlier iterations in the event of unforeseen problems or performance degradation.

4.7.8 Instruction and Assistance for User

- **User Training:** Teach operators, engineers, and decision-makers, among other end users, to understand model predictions, incorporate them into their workflows, and use the insights from the model to make well-informed decisions.
- **Record-keeping:** To maximize the utility of the deployed model, create thorough documentation and user guides that cover usage instructions, deployment methods, steps for troubleshooting, and best practices.

Semiconductor makers can efficiently implement predictive models in production settings, facilitating data-driven decision-making, streamlining manufacturing procedures, and improving yield and product quality by adhering to certain guidelines and best practices.

4.8 EVALUATION METRICS AND PERFORMANCE ANALYSIS

An essential part of evaluating the efficacy and dependability of predictive models in semiconductor production is performance analysis and evaluation measures. Here's a thorough examination of these elements.

4.8.1 Classification Metrics

In semiconductor manufacturing, tasks like defect detection or quality control and classification metrics are essential to assessing the efficacy of predictive frameworks. A high-level summary of the performance of the model is given by its accuracy – that is, the proportion of appropriately identified events. Precision gauges the model's resistance to false positives, which is crucial in circumstances where false alarms can be expensive. Recall, also known as sensitivity, quantifies the model's ability to detect every positive instance; this is particularly significant when reducing false negatives is essential. When there is an imbalance between precision and recall, the F1 score, which is their harmonic mean, balances them both and is helpful.

4.8.2 Regression Metrics

Regression metrics are useful in the semiconductor manufacturing industry for activities like process parameter estimation and yield rate prediction. The model's average size of errors is measured by the MAE, which provides a clear picture of how accurate the forecast is. The MSE, which punishes more severely for greater faults, measures the average of the squared disparities between the actual and projected values. Because it has the same units of expression as the target variable, the root mean squared error (RMSE), which is derived from MSE, is frequently employed for interpretability.

4.8.3 Extra Metrics

Receiver operating characteristic (ROC) curves and precision-recall curves are two additional tools that go beyond typical metrics and offer deeper insights into model performance, especially in binary classification tasks. The area under the ROC curve, or AUC-ROC, condenses this performance into a single scalar value, whereas ROC curves depict how the true positive rate and false positive rate are traded off over various thresholds for decisions. Contrarily, precision-recall curves provide a detailed understanding of the trade-offs between accuracy and recall, which is particularly helpful in unbalanced datasets.

4.8.4 Performance Analysis

To evaluate the reliability of a model and pinpoint areas in the need of development, performance analysis techniques, including cross-validation, model comparison, and error analysis are essential. Cross-validation provides insights into stability and robustness by splitting up the dataset into various subsets for training and testing, assisting in the estimation of a framework's capacity for generalization. By comparing models, practitioners can determine which model performs best on testing or validation data. Error analysis, on the other hand, explores the kinds and trends of mistakes the model makes, directing future iterations and enhancements.

4.9 CASE STUDIES AND APPLICATIONS

An explanation of case studies and the use of predictive modelling to increase yield in semiconductor production is provided here.

4.9.1 Defect Identification and Classification

In the semiconductor manufacturing process, predictive modelling approaches are extensively utilized for the purposes of defect identification and classification.

Predictive models are able to recognize and categorize several kinds of defects on semiconductor wafers or chips, such as particles, scratches, or unusual patterns, by examining sensor data, metrology results, and inspection images. To understand intricate patterns and features suggestive of flaws, these models make use of deep learning architectures like CNNs. This allows for proactive quality management and minimizes yield loss.

4.9.2 Process Control and Optimization

Predictive modelling is essential for regulating important parameters and streamlining manufacturing processes to improve yield in the production of semiconductors. Predictive models can determine the best process conditions, anticipate equipment failures or maintenance requirements, and suggest changes to increase yield and productivity by examining past process data and equipment logs. These models use methods like reinforcement learning and time-series analysis to dynamically optimize process parameters and adjust to shifting production settings.

4.9.3 Predictive Maintenance

It is an essential use of predictive modelling in the semiconductor industry that aims to maximize equipment uptime and minimize downtime by anticipating equipment breakdowns before they happen. Predictive models can anticipate equipment failures, identify abnormal behaviour, and plan maintenance tasks ahead of time by evaluating sensor data, equipment logs, and maintenance records. By using machine learning methods like RNNs or support vector machines to recognize trends suggestive of equipment malfunction or degradation, these models enable predictive maintenance procedures and minimize unscheduled downtime.

4.9.4 Yield Prediction and Optimization

The prediction and optimization of yield are fundamental goals in the semiconductor manufacturing process, as even minor enhancements in yield can offer substantial economic benefits. To estimate yield rates and pinpoint areas for improvement, predictive models examine a variety of yield-influencing factors, such as material qualities, environmental factors, and process parameters. To estimate yield dependencies, optimize process parameters, and reduce variability, these models make use of statistical methodologies, machine learning algorithms, and optimization approaches. This eventually improves total yield and profitability.

4.9.5 Fault Detection and Root Cause Analysis

Predictive modelling plays a key role in semiconductor manufacturing's fault detection and root cause analysis processes, assisting in the identification and mitigation of problems that affect yield and product quality. Predictive models scan sensor data, equipment logs, and historical performance data to identify anomalies and their underlying causes and suggest remedial measures. These models make use of expert systems, causal inference methods, and anomaly detection algorithms to identify fault locations, comprehend their underlying causes, and apply focused fixes that enhance process yield and stability.

These case studies and applications show how flexible and useful predictive modelling approaches are for increasing yield, streamlining workflows, and guaranteeing product quality in the semiconductor manufacturing industry. Chipmakers can get fresh insights, promote operational excellence, and retain a competitive edge in the market by utilizing advanced analytics, machine learning algorithms, and domain expertise.

4.10 FUTURE DIRECTIONS AND EMERGING TRENDS

Examining the probable directions and trends in predictive modelling for semiconductor manufacturing yield enhancement provides insightful information about future developments and advancements that could shape the industry. Here's a thorough rundown.

4.10.1 Advanced Deep Learning Architectures

It is projected that future advancements in deep learning architectures will completely transform semiconductor manufacturing's predictive modelling. Further investigation into sophisticated structures, such as capsule networks, attention mechanisms, and graph neural networks, has the potential to improve the efficiency with which complicated relationships and dependencies in manufacturing data are captured. More reliable and accurate predictive models may be made possible by these designs, especially for jobs involving high-dimensional and diverse data sources.

4.10.2 Explainable AI

To improve model interpretability and transparency, explainable AI (XAI) techniques are becoming more and more necessary, as AI-driven decision-making is used more often in semiconductor manufacturing. To help stakeholders comprehend, validate, and have faith in predictive model outputs, future research endeavours will concentrate on creating XAI techniques that offer insights into how predictive models make judgements. Attention mechanisms, feature importance analysis, and model-agnostic explanation techniques are some of the key techniques that will help advance XAI in semiconductor manufacturing.

4.10.3 Domain-Specific Data Augmentation

A crucial field of study will be the development of domain-specific data augmentation methods suited to the semiconductor manufacturing industry. These methods seek to alleviate data scarcity problems, improve model generalization, and produce synthetic data that emulates actual manufacturing settings. In the future, it might be possible to produce representative and diverse datasets for training predictive models by utilizing domain-specific augmentation techniques, variational autoencoders, and generative adversarial networks.

4.10.4 Multimodal Data Fusion

In semiconductor production prediction modelling, the incorporation of multimodal data sources, like sensor, image, and textual records, will become more common. Subsequent investigations will concentrate on formulating strategies for multimodal data fusion, which will provide thorough examination and simulation of intricate manufacturing procedures. Methods like transfer learning, fusion-based architectures, and multitask learning will make it easier to integrate disparate data sources and improve predictive modelling abilities.

4.10.5 Federated Learning and Edge Computing

These two concepts will become ground-breaking tools for implementing predictive models in semiconductor manufacturing settings. Latency is lowered using edge computing, which also improves reactivity by facilitating real-time inference and decision-making right on the manufacturing floor. Federated learning keeps data security and privacy through making collaborative model training responsible across dispersed edge devices without exchanging sensitive data. Decentralized predictive modelling in semiconductor production will be made possible by future developments in communication protocols, federated learning algorithms, and edge computing infrastructure.

4.10.6 Digital Twin Technologies

In semiconductor production, digital twin technologies will be essential to enable simulation-driven optimization and predictive modelling. Digital twins offer a platform for real-time production system monitoring, analysis, and improvement. They are virtual copies of actual manufacturing processes. Predictive models will be integrated into digital twin settings in the future, allowing for proactive decision-making for yield enhancement, scenario analysis, and predictive maintenance.

4.10.7 Applications of Quantum Computing

Predictive modelling in semiconductor manufacturing is a computationally demanding process that quantum computing can revolutionize. The potential for exponentially faster calculations is presented by quantum algorithms, such as those used in quantum machine learning and quantum optimization, which should make model training, optimization, and inference more effective. Subsequent investigations will examine the practicability of quantum computing implementations in semiconductor fabrication, tackling obstacles associated with hardware scalability, error reduction, and algorithm creation.

4.10.8 Ethical and Regulatory Considerations

These issues will gain prominence as predictive modelling spreads throughout the semiconductor production industry. The establishment of best practices, standards, and guidelines for data governance, algorithmic transparency, and responsible AI deployment in semiconductor production will be the main goals of future projects. Encouraging the proper use of predictive modelling technology and addressing ethical concerns will need the cooperation of academic institutions, regulatory authorities, and industrial players.

Semiconductor makers can fully utilize predictive modelling to improve yield, streamline operations, and spur industry innovation by embracing these new directions and trends. Predictive modelling capabilities for semiconductor manufacturing will undergo revolutionary changes because of ongoing research, teamwork, and technology breakthroughs.

4.11 CONCLUSION

In summary, yield enhancement predictions are becoming more accurate, signalling a revolution in semiconductor production. Predictive modelling has become a key part of process optimization, product quality improvement, and operational efficiency in semiconductor production facilities thanks to the combination of cutting-edge technologies, creative approaches, and domain expertise.

This investigation has shed light on the critical role that predictive modelling plays in many different aspects of semiconductor manufacturing. Predictive models have proven their ability to decipher complex patterns, extract actionable insights, and equip decision-makers with the foresight required to navigate the complexities of contemporary semiconductor production. These insights have been applied to defect detection and classification, process optimization, predictive maintenance, and yield prediction, among other applications. Furthermore, because of the small number of data points, deep learning models are not examined in this work; nonetheless, they must be incorporated in subsequent studies [7]. It is clear that predictive models' trajectory in semiconductor manufacturing is set to continue evolving as we look to the future. Emerging trends that will change the game and pave the way for innovation and discovery include XAI, edge computing, and domain-specific data augmentation.

But in the middle of all the excitement surrounding technical advancement, it's critical to emphasize how crucial ethical issues, legal compliance, and appropriate AI deployment are. As predictive modelling spreads, maintaining ethical norms, protecting data privacy, and guaranteeing algorithmic openness will be critical to building cooperation, trust, and a long-lasting ecosystem of innovation and advancement.

The journey of predictive modelling for yield enhancement in semiconductor manufacturing is essentially a story of three people who have never stopped pursuing perfection. Predictive modelling is like a lighthouse guiding us through the maze of the semiconductor business, showing us the way to a future where quality, efficiency, and precision work together to push the boundaries of what is possible.

REFERENCES

1. Melhem, M.; Ananou, B.; Ouladsine, M.; Combal, M.; Pinaton, J. Product Quality Prediction Using Alarm Data: Application to the Semiconductor Manufacturing Process. In *2017 25th Mediterranean Conference on Control and Automation (MED)*, 2017, pp. 1332–1338. https://doi.org/10.1109/MED.2017.7984303.
2. Kovacs, I.; Topa, M.; Buzo, A.; Pelz, G. An Accurate Yield Estimation Approach for Multivariate Non-Normal Data in Semiconductor Quality Analysis. In *Proceedings of the 2017 14th International Conference on Synthesis, Modelling, Analysis and Simulation Methods and Applications to Circuit Design (SMACD), IEEE*, Giardini Naxos, Italy, 12 June 2017, pp. 1–4.
3. Lee, D.-H.; Yang, J.-K.; Lee, C.-H.; Kim, K.-J. A Data-Driven Approach to Selection Process Steps in the Semiconductor Manufacturing Process Considering Missing and Imbalanced Data. *J. Manuf. Syst.* 2019, 54, 146–156.
4. Sarker, Iqbal H. Data Science and Analytics: An Overview from Data-Driven Smart Computing, Decision-Making and Applications Perspective. *SN Comput. Sci.* 2021, 20, 21.
5. Sarker, I.H.; Hoque, M.M.; Uddin, M.K.; Alsanoosy, T. Mobile Data Science and Intelligent Apps: Concepts, AI-Based Modelling and Research Directions. *Mob. Netw. Appl.* 2020, 26, 285–303.
6. Stephan, G.; Pascal, H.; Andreas, A. Knowledge Representation and Ontologies. *Semant. Web Serv. Concepts Technol. Appl.* 2007, 20, 51–105.
7. Dou, Z.; Sun, Y.; Zhang, Y.; Wang, T.; Wu, C.; Fan, S. Regional Manufacturing Industry Demand Forecasting: A Deep Learning Approach. *Applli. Sci.* 2021, 11, 6199.

5 Deep Learning for Image Classification in Semiconductor Inspection

Vanshika Jain, Rashmi Gupta, Neeraj Gupta, Prashant Kumar and Rajat Butola

5.1 INTRODUCTION

Semiconductors represent the backbone of modern technology and are critical components in manufacturing technologies. So ensuring their perfect production is vital – and even more so in light of the prevailing semiconductor shortages. Ideally, the defect detection technology for semiconductor wafers should take into account the wafer as a whole. When a defect is found, it is usually recorded on a wafer bin map to discover how many similar defects exist and whether they are concentrated in certain areas of the entire surface of the wafer. This method effectively identifies systemic defects and allows premature elimination of defective products from the manufacturing process. However, it does not provide detailed enough information for necessary distinction between several types of defects.

The main purpose of this chapter is to expand on the task of simply determining the location of defects and focusing more on the classification of defects according to their type. This endeavour is of foremost importance, as it helps in the determination of errors that may not make the product completely useless. Since defects can be grouped, it becomes easy to respond appropriately to recover good chips that are still saleable, hence making the best use of resources and cutting down on losses. To this end, semiconductor die images are analysed. These images are taken from high-resolution photos of individual chips on a wafer and give a microscopic vision of the semiconductor field and an idea about the type and extents of defects. Figure 5.1 shows the defect categorization in semiconductor die images.

The manufacturing of semiconductor wafers is a complex and complicated process since it is an activity that is made up of so many interrelated stages. Starting right from the growth of silicon crystals to the micro-fabrication of transistor structures, each and every process calls for discipline and control to meet the set quality requirements. Additionally, the process of manufacturing semiconductors is complex by all standards, and when combined with the criticality of the parts that make up electronic devices, it becomes evident why a broad-ranging approach to quality management is needed.

DOI: 10.1201/9781003508304-5

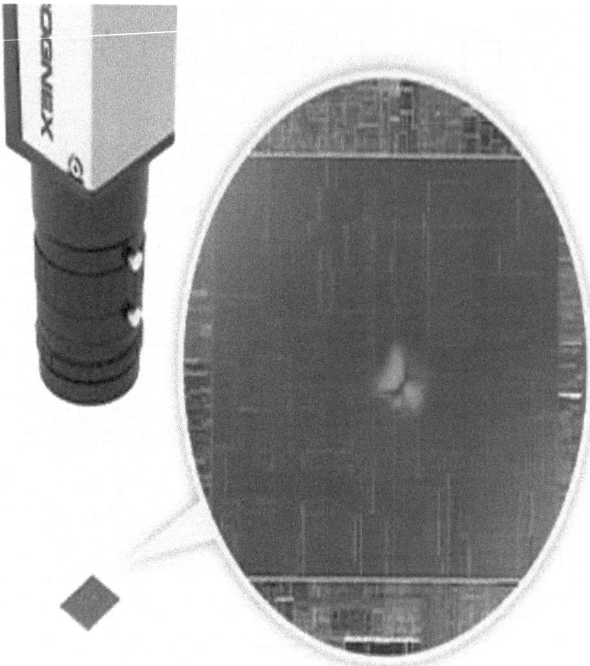

FIGURE 5.1 Defect categorization using semiconductor die images.

Source: Copyright © Cognex Corporation, "Die surface inspection." Cognex. www.cognex.com/industries/electronics/semiconductors/die-surface-inspection.

Due to the varying demands in the production of semiconductors, a variety of inspection methods is used at all stages of production. These techniques can be classified into a wide range of strategies that includes macro analysis and microanalysis of the subject. For instance, electrical testing and microwave analysis provide information on functional performance, while optical cameras and scanning electron microscopes (SEMs) help in analysing the nanoscale features. However, due to the fact that these techniques are diverse, there are certain issues that arise from the aspect of defect identification and classification and therefore require a united approach that will harness the strengths of the techniques while avoiding their shortcomings. Defect detection operations can be enhanced with the help of data analytics and machine learning algorithms as well as methods based on artificial intelligence (AI); this will help maintain high-quality control throughout the semiconductor manufacturing process. Figure 5.2 shows the defects at different stages of production.

5.2 EVOLUTION OF SEMICONDUCTOR MANUFACTURING AND INSPECTION

The methods of manufacturing and inspecting semiconductors can be described as gradual progress with constant development of new techniques. Such a

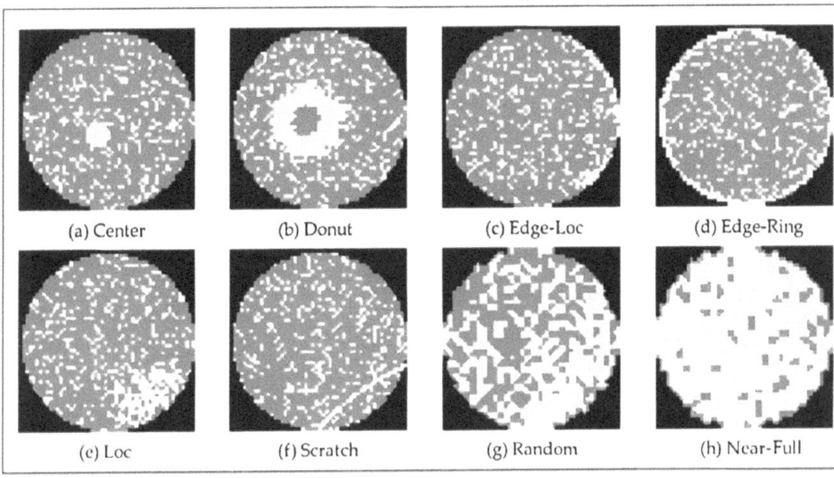

FIGURE 5.2 Showing the defects in semiconductor wafers during production.

Source: Copyright © M. Saqlain, Q. Abbas, and J. Lee, "A deep convolutional neural network for wafer defect identification on an imbalanced dataset in semiconductor manufacturing processes," ***IEEE Trans. Semicond. Manuf.*****, vol. 33, no. 3, pp. 436–444, 2020. https://doi.org/10.1109/TSM.2020.2994357.**

progression has been helpful in the development of the semiconductor industry by creating a pathway to generate more complex chips. Here is a detailed overview of the key milestones and advancements that have shaped the current state of the industry:

1. **Early Semiconductor Fabrication Techniques**
   - When semiconductor manufacturing was in its infancy, it was mostly a manual process and primitive. Transistors and early integrated circuits (ICs) were fabricated using techniques like photolithography and chemical etching.
   - Copper was commonly used as a semiconductor material, but silicon was favoured because of its abundance and favourable semiconductor characteristics.
2. **Introduction of Cleanroom Environments**
   - Cleanroom technology became formalized in the 1960s, although this was the period when it was first introduced in semiconductor production. Cleanrooms are environments that are designed to have a low level of airborne particles, as these are important in the protection of semiconductor materials.
   - Cleanroom classifications are still dynamic since there are continual improvements in filtration, ventilation, and contamination control systems.
3. **Advancements in Photolithography**
   - Photolithography became the key technology process in the field of semiconductor devices on silicon wafers. At the beginning of photolithography, the wafers were coated with a photosensitive material known

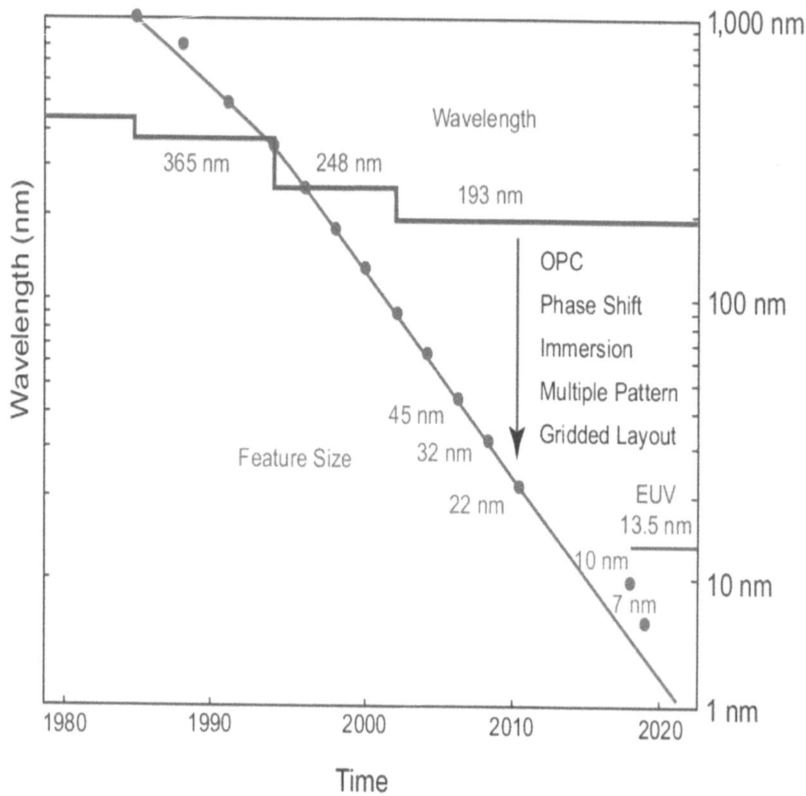

FIGURE 5.3 Historical progression of IC feature size and photolithography technologies.

Source: Copyright © Newport Corporation, "Photolithography overview." ***Newport.*** **https://www.newport.com/n/photolithography-overview.**

as photoresist, exposed to light through a mask, and then chemically etched to replicate the pattern.

- In later years, improvements in the photolithography process like deep ultraviolet and extreme ultraviolet lithography allowed for the fabrication of fine features and more transistors. Figure 5.3 describes the historical progression of IC feature size and photolithography technologies.

4. **Introduction of Process Integration**
   - In the 1970s, the idea of combining different manufacturing steps into one smooth process started to become important. This method helps make more advanced electronic parts by putting together many complex pieces.
   - Combining steps like putting materials on the surface, cutting away parts, and adding specific materials is part of process integration. This helps create detailed parts and connections on the silicon pieces used in electronics.

5. **Introduction of Wafer Steppers and Scanners**
   - Wafer steppers and scanners changed how semiconductors are made by allowing the whole surface of wafers to be exposed to light at once, which speeds up the process and makes more chips.
   - Steppers and scanners use special lenses and sophisticated systems to place circuit designs accurately and consistently onto silicon wafers.
6. **Emergence of Semiconductor Metrology and Inspection Tools**
   - As the size of parts in semiconductor devices gets smaller, it's very important to have accurate measuring and checking tools. Tools like SEMs, atomic force microscopes (AFMs), and optical inspection systems are now essential for looking at the tiny parts of semiconductors and finding any small mistakes.
   - These tools help companies that make semiconductors check the size of their devices, find any mistakes, and ensure everything is made correctly during the whole production process.
7. **Integration of Advanced Process Control Systems**
   - Advanced process control (APC) systems use data analytics and machine learning algorithms to check, analyse, and optimize semiconductor fabrication processes in real time. APC systems enable semiconductor manufacturers to achieve consistent product quality and maximize yield.
   - APC systems incorporate sensors, actuators, and feedback mechanisms to adjust process parameters and maintain optimal performance.
8. **Rise of Computational Lithography**
   - Computational lithography has emerged as a critical enabler of advanced semiconductor patterning techniques. By leveraging computational models and algorithms, semiconductor manufacturers can simulate and optimize complex lithographic processes to achieve higher resolution and pattern fidelity.
   - Computational lithography is critical for designing and manufacturing next-generation semiconductor devices with reduced feature sizes and increased transistor density.
9. **Introduction of Defect Review and Classification Systems**
   - As semiconductor devices become more complicated, the need for better defect review and classification systems has grown. These systems use machine learning algorithms and image analysis techniques to accurately detect, categorize, and analyse faults on semiconductor wafers.
   - Defect review and classification systems enable semiconductor manufacturers to rapidly diagnose faults, optimize process parameters, and improve product quality.
10. **Integration of Industry 4.0 Technologies**
    - Industry 4.0 principles, such as automated processes, connectivity, and data-based decision-making, are making semiconductor manufacturing more efficient and agile. Advanced robotics, internet of things devices, and cloud-based analytics platforms are redefining the way semiconductor manufacturing facilities function.

- Industry 4.0 technologies enable predictive maintenance, real-time process optimization, and adaptive manufacturing strategies, enhancing productivity, flexibility, and quality in semiconductor manufacturing.

5.3 CHALLENGES IN SEMICONDUCTOR INSPECTION

Semiconductor inspection faces several challenges, which are closely related to companies' drive for innovation and the shrinking sizes of semiconductors. The foremost challenge is the rise in semiconductor designs' complexities. Each leap in semiconductor generation leads to ICs turning into increasingly complex structures, characterized by many more layers, finer features sizes, and tighter tolerances. Consequently, the tangled web of intricacy presents limitations that ambitious detection schemes must cut through to accurately capture any faults or deviations along multiple layers and interwoven parts [1].

Furthermore, another layer of complexity is added during the inspection process due to the relentless demand for high throughput in manufacturing. A semiconductor facility runs at breakneck speed, driven by the insatiable appetite for semiconductors. As such, inspection systems should be capable of dealing with large volumes of wafers or chips with ease and accuracy. Classical investigation techniques, which were sufficient when industries ran at a slower pace, cannot keep up with today's fast-moving production lines. Thus, it is necessary that automated inspection tools are developed.

Moreover, for semiconductor manufacturers, achieving even higher yields while maintaining high quality will continue to be one of the most challenging aspects of fabricating products with the expected performance and reliability required by critical computing and communication applications. The need for improved yields is a key factor for keeping semiconductor enterprises highly competitive and for maintaining healthy profitability in a hypercompetitive marketplace. The need is also growing more challenging as semiconductor devices become more complex. Semiconductor inspection serves as a key enabler for yield improvement through defect identification and characterization that can potentially impact product performance and reliability. The need to increase yield objectives while maintaining high quality requires the ability to develop powerful defect characterization and classification schemes that benefit from machine learning algorithms and AI to detect defects with greater precision while simplifying the process of optimizing manufacturing operations.

Also, APC systems use advanced techniques to manipulate systems, which adds an extra layer of complexity in semiconductor inspection. APC devices on the leading edge of true-time monitoring and control require symbiotic integration with inspection facts to initiate modifications for yield optimization. Strong facts, analytics competencies, and smooth interoperability between the semiconductor inspection device and manufacturing gadget are essential to realize this symbiotic integration, and complex modelling and predictive analytics strategies are critical to anticipate and prevent capability defects before the failures manifest, all while semiconductor devices fulfil growing marketplace demands amid pervasive technical changes [2].

5.4 LITERATURE REVIEW ON CNN-BASED DEEP LEARNING MODELS

The development and applications of several types of artificial neural networks have been in both research and production. Nevertheless, convolutional neural networks (CNNs) have been in the limelight for the remarkable results they have achieved in the field of image recognition, especially on datasets such as the MNIST database, CIFAR-10, and ImageNet [3]. The capability of CNNs to model the grid-like nature of images and smaller computation requirements than conventional neural networks, owing to their spatially organized neurons and pooling mechanisms, account for their superior performance in this application. What's more, CNNs showcase a remarkable capacity to recognize the differences between a huge number of classes. In fact, CNNs represent a kind of end-to-end learning that decreases the human contribution requirement. Their design fuses the processes of feature

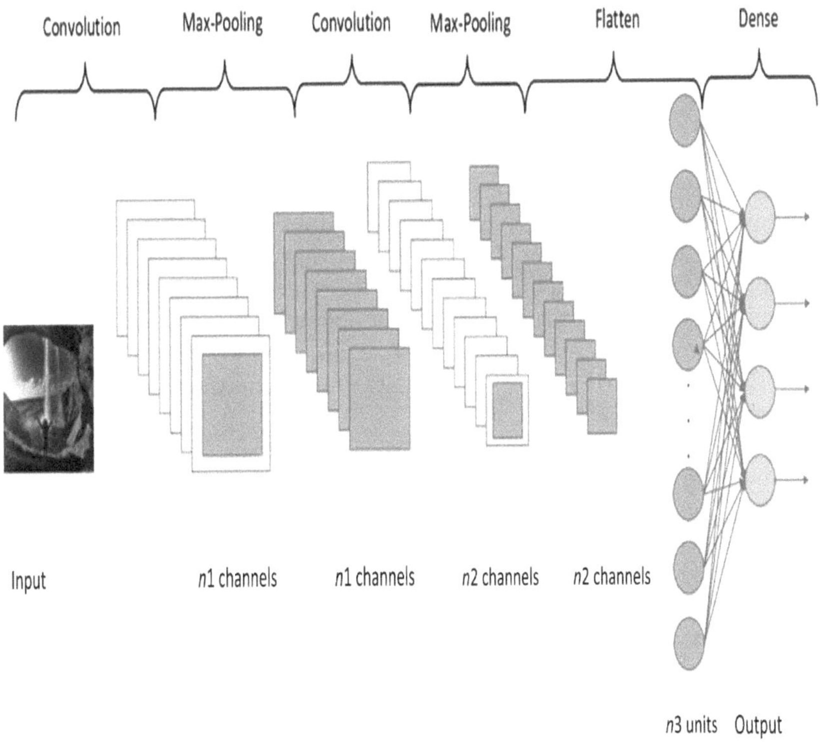

FIGURE 5.4 Convolutional neural network architecture.

Source: Copyright © A. Harb, "Convolutional neural network tutorial for beginners," ***freeCodeCamp*****, 28 December 2020. https://www.freecodecamp.org/news/convolutional-neural-network-tutorial-for-beginners/.**

extraction and classification together, thereby making them highly appropriate for efficient image recognition tasks [4]. CNNs have four main layers: convolution, rectified linear unit (ReLU) activation, pooling, and connected layers. After several rounds of convolution and pooling, CNNs use connected layers to reason at a prominent level, working out class scores for the data that comes in. In cases with multiple classes, they use the SoftMax function. This function gives back probabilities for each class of the input, making sure all the probabilities add up to 1. Figure 5.4 highlights the CNN architecture.

5.5 CNN-BASED CLASSIFICATION ARCHITECTURE

In the past, CNNs faced big challenges because they needed massive amounts of labelled data, which led to high processing costs. This issue got better when people created large well-labelled datasets like ImageNet, CIFAR-10, CIFAR-100, and the MNIST database. Also, improvements in parallel computing with graphics processing units (GPUs) and big distributed clusters lowered computing limits. This made it possible to better handle these massive datasets.

The ImageNet large-scale visual recognition challenge (ILSVRC) had an enormous impact on pushing forward deep neural networks for computer vision. ILSVRC made it simpler to build and evaluate more complex neural network structures by providing a standard against which to measure different image classification systems.

The ImageNet challenge led to two key image classification systems: the VGG Network (VGG-Net) and the Residual Network (ResNet). Simonyan and Zisserman created VGG-Net in 2014, which had a significant impact on exploring deeper network structures. VGG-Net used tiny 3 × 3 convolutional filters, which helped the network spot small details in images while keeping the number of parameters under control. This design let the network grow to 16–19 weight layers, achieving a high accuracy of 7.3% at the ILSVRC 2014. The network's increased depth improved classification performance, setting a brand-new standard for CNN architecture [5].

ResNet, which Microsoft Research Asia introduced in 2015, brought about a momentous change with its game-changing 152-layer network structure. The main breakthrough of ResNet was residual learning. This involved learning residual functions in relation to input layers instead of just learning the intended underlying mapping. This method made it easier to optimize very deep networks by dealing with the degradation problem. This problem happens when adding more layers results in worse performance. ResNet's design took full advantage of the much greater depth. This led to better accuracy, and it won the ILSVRC 2015 with a low error rate of 3.6% [6].

Building on the foundation set by these ground-breaking works, our study aims to use transfer learning methods and adapt these top-performing network designs for our Deep Learning for Anomaly Detection and Classification (DLADC) project. By tapping into the unique features of VGG-Net and ResNet, we hope to expand the limits of anomaly detection and classification, boosting accuracy and effectiveness in spotting and sorting abnormalities in various datasets.

5.6 SUITABILITY OF CNN FOR IMAGE CLASSIFICATION

Research on using CNNs to check and find defects in semiconductors keeps expanding. CNNs have an edge over regular machine learning methods. They can learn complex patterns straight from the images of semiconductors without needing humans to step in or create unique features. This ability proves helpful in making chips and wafers. It allows manufacturers to spot flaws, which helps them keep their products top-notch and reliable.

CNNs have a knack for spotting complex patterns and structures in semiconductor images. They can pick out features using convolutional and pooling layers at distinct levels of detail. Thanks to the grid-like setup in semiconductor pictures, CNNs can spot tiny flaws like particles, scratches, or pattern changes that might point to manufacturing issues. What's more, CNNs are flexible in defect detection tasks. They can adapt to differences in semiconductor pictures caused by things like various fabrication methods, materials, and equipment settings.

CNNs are fit for semiconductor inspection. They can handle tons of visual info and work well on new samples. Training CNNs on various defects helps them spot more issues. There are labelled datasets for semiconductor inspection, like SEMI's (Semiconductor Equipment and Materials International) Standards and Guidelines for Defects in Semiconductor Manufacturing. Also, better hardware acceleration tech such as GPUs and special accelerators, allows for smooth integration of CNN-based inspection systems in chip factories. This tech boosts output and scalability too.

To wrap up, CNNs show great promise for inspecting semiconductors. They offer automated, accurate, and scalable ways to spot defects and control quality in semiconductor production. As tech in this area keeps advancing, CNNs are set to play a key role in making semiconductor production more effective and reliable.

5.7 DATA COLLECTION FOR INSPECTION

1. **Wafer Images**
    - **Manufacturing Environment:** People take wafer images in semiconductor fabrication plants (fabs). These places have strict controls, like cleanrooms, to keep contamination levels low and make sure the data stays good.
    - **Automated Imaging Systems:** Automated imaging systems with high-resolution cameras take wafer pictures. These systems can be part of the semiconductor manufacturing equipment or exist as separate inspection stations.
    - **Sequential Imaging:** Cameras snap wafer images one after another as the wafer goes through different fabrication steps, like deposition, lithography, etching, and annealing. This step-by-step imaging helps manufacturers to keep an eye on how the wafer's surface changes and spot any problems or flaws right away.
    - **Metadata Annotation:** Manufacturers add metadata such as wafer ID, lot number, fabrication step, and timestamp to the wafer images. This makes it easier to track and trace the wafers.

FIGURE 5.5 Sample silicon wafer.

Source: Copyright © "Why do silicon wafers look rainbow colored?," Stack Exchange, 14 July 2021. https://electronics.stackexchange.com/questions/573975/why-do-silicon-wafers-look-rainbow-colored.

Figure 5.5 depicts the image of a sample silicon wafer.

2. **Die Images**
   - **Die Extraction:** Image processing techniques like segmentation and cropping help to extract die images from wafer images. These methods find and isolate each individual chip (die) on the wafer to generate die-level images for inspection.
   - **High-Resolution Imaging:** Die images, just like wafer images, are taken with high-resolution cameras. This ensures a detailed look at each chip. The high-resolution pictures help spot tiny flaws and odd features that could affect how well the chip works.
   - **Multiple Views:** Cameras might snap die images from different angles (like from above or at a slant). This captures various surface features and helps to analyse defects.
3. **Defect Images**
   - **Defect Localization:** Automated algorithms or human operators find defects in wafer or die images. Once they spot these flaws, they mark the areas with defects for a closer look later.
   - **High-Magnification Imaging:** Images of defects are taken at high magnification to show the defects in detail. This often involves using special microscopes, like SEM or AFM, to see things as small as a nanometre.

- **Defect Classification:** Images of defects are sorted by type, size, shape, and how bad they are. This helps makers focus on the most important ways to fix defects and make their processes better to stop similar defects from happening again.

4. **Metadata Annotation**
   - **Semantic Annotation:** Along with marking defects on images, extra info like the type of defect, its size, where it is, and how bad it is gets added to the image data. This provides more background information when looking at the defects.
   - **Database Integration:** Annotated defect images and related info go into central databases or data management systems. This helps people find, analyse, and make decisions about data during chipmaking.

5.8 DATA PREPROCESSING

Getting semiconductor pictures ready for machine learning analysis is key. People often use these methods to make image data more accurate and useful:

1. **Normalization:** Normalization converts the value of each pixel of images to a common scale, usually [0, 1] or [−1, 1]. It keeps all pixel intensities the same across images, so the machine learning model can spot more patterns and features in the data.
2. **Augmentation:** Image augmentation techniques create more training examples from source images by making changes like rotating, flipping, scaling, or cropping. This process has an impact on the range and strength of training data, reducing the risk of overfitting, and improving the model's ability to generalize.
3. **Noise Reduction:** Noise reduction methods clean up or lessen unwanted elements and distortions in images, which can happen because of things like sensor noise, dust, or image compression. Common approaches include smoothing filters such as Gaussian blur or median filtering, which help keep important image features while cutting down on noise.

Using these preprocessing methods has an impact on how well semiconductor images are prepared to analyse with machine learning models. This ensures that the models can learn and pick out meaningful patterns and features from the data accurately, which leads to more reliable and efficient systems for inspecting semiconductors.

5.9 MODEL ARCHITECTURE

1. **LeNet:** Yann LeCun and his team came up with the LeNet in the 1990s. It's one of the first CNN designs. LeNet has different layers, like convolutional layers, max-pooling layers, and fully connected layers. It's not as complex as newer designs, but LeNet still does a great job with basic picture sorting tasks, like those in semiconductor checks. Its straightforward layout opened doors for more advanced CNNs. This showed how powerful deep learning can be for working with images.

2. **AlexNet:** AlexNet, which Alex Krizhevsky and his team came up with in 2012, marked a big step forward in CNN design. The structure has several convolutional layers, with max-pooling layers right after them, and connected layers at the end. When AlexNet won the ILSVRC, it showed how good deep learning is for sorting pictures. People have used this model in many areas, including checking semiconductors, because it can handle complex picture data and still sort things.
3. **VGG:** Karen Simonyan and Andrew Zisserman came up with the VGG network in 2014. This network stands out because of its uniform structure, which uses small (3 × 3) convolutional filters and max-pooling layers [7]. The uniform design of VGG networks makes them simple yet good at classifying images. The architecture takes a direct approach to making the network deeper (up to 19 layers) while keeping the computing needs in check. This has proven helpful in sorting semiconductor images where being clear and exact matters.
4. **ResNet:** Kaiming introduced ResNet. In 2015, he and his team came up with a ground-breaking method for deep learning called residual learning. This involves creating shortcut (or skip) connections that skip one or more layers. These connections allow very deep networks, often with hundreds of layers, to be trained by solving the vanishing gradient problem, which often hinders deep network training. ResNet has reached top-notch performance in many image classification tasks, and people praise it for its strength and effectiveness. Its ability to keep performing well as it gets deeper makes it a viable choice for the complex job of inspecting semiconductors.

5.10 MODEL TRAINING

1. **Data Splitting**
   - **Purpose:** Splitting data plays a key role in training and checking how well the CNN model works. It involves breaking up the dataset into three different parts: sets to train, validate, and test.
   - **Training Set:** The set for training makes up a big part of the dataset and helps teach the CNN model. As it trains, the model learns to spot patterns and features in the pictures of semiconductors along with their tags.
   - **Validation Set:** This set helps to adjust settings and keep an eye on how the model does during training. By checking how the model performs on the validation data regularly, we can make changes to stop it from fitting too much or too little to the data.
   - **Testing Set:** The testing set helps to check how well the trained model performs. It gives a fair assessment of the model's ability to work with new data, ensuring it does well in real-world cases.
2. **Hyperparameter Tuning**
   - **Learning Rate:** The learning rate decides how big the steps are when updating parameters during training. It is a key setting that has a big effect on how fast the model improves and how well it works overall. We can use learning rate methods and adaptive learning rate methods (like Adam) to change the learning rate as training goes on.

- **Batch Size:** Batch size decides how many samples are looked at in each training round. Bigger batches help reach the goal faster, but they need more memory and computing power. Methods like batch normalization and gradient accumulation can help keep training steady when using large batches.
- **Regularization Parameters:** Regularization tricks like L1 and L2 regularization or dropout stop overfitting. They do this by keeping parameter values in check or turning off units during training. The strength of regularization depends on hyperparameters. These need to be adjusted to make the model work its best.
- **Hyperparameter Search:** This involves looking into different hyperparameter combinations to make the model work better. You can use methods like grid search, random search, and Bayesian optimization to check out various hyperparameter options and find the best ones.

3. **Optimization Algorithms**
   - **Stochastic Gradient Descent:** Stochastic gradient descent (SGD) is a way to train machine learning models in deep neural networks. It is different from regular gradient descent methods. SGD tweaks the model's settings for individual data points or small groups rather than looking at all the data at once. This approach makes calculations faster and helps the model learn quicker. SGD works well with big datasets where it's not possible to process everything in one go.
   - **Adam:** Adam is a learning rate optimization algorithm that is open to input. It uses momentum and RMSprop. It changes the training speed for each parameter based on the first and second moments of the gradients. This leads to quicker convergence and better standardization.
   - **RMSprop:** RMSprop is another algorithm that adapts the learning rate. It uses a moving average of squared gradients to work out the training rate for each parameter in the algorithm. It solves the problem of gradients that vanish or burst by changing learning rates on the fly based on the most recent gradient history.
   - **Loss Function:** The loss function you use depends on what you are trying to do with your semiconductor inspection. If you are sorting things into many groups, cross-entropy for multiple categories is a go-to choice. For yes-or-no decisions, binary cross-entropy works well. When you need to predict numbers, mean squared error can do the job.

By looking at how to split data, adjust settings, and pick the right math tricks, companies can build and improve CNN models that work well for checking semiconductors. This helps them spot defects and keep quality high when making these tiny electronic parts.

5.11 MODEL TESTING

1. **Accuracy:** Accuracy is a basic statistic that shows the percentage of grouped items in a dataset. It gives a full look at how right the model is by showing how many predictions match the real labels. For checking semiconductors, accuracy is a key sign of how well the model spots flaws or odd things in

semiconductor images. But accuracy alone might not tell the whole story with uneven datasets, where one group is much bigger than others. So while accuracy gives important info about how useful the model is, it's best to use it along with other measures to get a better picture of how well the model works.

2. **Precision:** Precision measures how well the model avoids false positives by calculating the ratio of true positive predictions to all positive predictions the model makes. This metric is key in semiconductor inspection because it shows how well the model can spot real defects without flagging good parts as bad. High precision means the model raises very few false alarms, which is crucial to cut down on unnecessary actions or follow-up steps caused by these false positives. Precision serves as a vital performance metric that manufacturers can use to evaluate how reliable and trustworthy defect detection systems are in semiconductor production environments [8].
3. **Recall (Sensitivity):** Recall, also called true positive rate, evaluates how well the model can show all positive cases. It does this by comparing the share of true positive predictions to all actual positive instances in the dataset. In chip testing, recall plays a key role in ensuring thorough defect detection. It shows how well the model can spot all faulty parts without missing any. High recall means the model identifies most positive cases, which lowers the chance of unnoticed flaws getting through the inspection. Makers focus on recall when missing flaws could have profound consequences. This highlights how important it is to have strict defect detection in chip production lines.
4. **F1 Score:** The F1 score strikes a balance between precision and recall by calculating their harmonic mean. It offers a thorough assessment of how well a model performs by taking into account both false positives and false negatives. In the field of semiconductor inspection, the F1 score serves as a complete evaluation metric that sheds light on the connection between precision and recall. A high F1 score suggests that the model achieves an optimal blend of precision and recall, cutting down on false alarms while capturing all positive instances. Manufacturers use the F1 score to examine the overall effectiveness and reliability of the defect detection system, making sure they evaluate its performance across different operating conditions.
5. **Area under the ROC Curve:** The area under the receiver operating characteristic (ROC) curve, or AUC-ROC, assesses how well a model can tell different classes apart. It does this by showing the true positive rate against the false positive rate at various threshold levels. A higher AUC-ROC score means the model can better distinguish between classes, with higher true positive rates and lower false positive rates at different decision points. In semiconductor inspection, the AUC-ROC plays a key role to gauge how well the model can spot the difference between faulty and good parts. Manufacturers rely on the AUC-ROC to check the model's overall sorting accuracy and its ability to make sound choices when looking at semiconductor images.
6. **Confusion Matrix:** A confusion matrix shows how well a model predicts compared to actual labels. It displays true positives, false positives, true negatives, and false negatives. This gives us key insights into the model's performance across different classes, letting us take a closer look at what it

does well and where it falls short. In semiconductor inspection, a confusion matrix helps makers spot specific areas where the model shines or struggles. This allows them to zero in on improvements. By looking at the confusion matrix, manufacturers can better understand how the model predicts and make smarter choices to boost its ability to spot and sort defects.

Each of these evaluation metrics gives us different insights into how well deep learning models work for checking semiconductors. This helps manufacturers make their defect detection systems better and produce top-notch semiconductor devices.

5.12 RESULTS AND DISCUSSION

In this chapter, we took a close look at how well CNN models perform when assessing semiconductor images. We ran careful tests to evaluate different CNN designs, such as AlexNet, VGG, and ResNet, using semiconductor image datasets from factories. We chose these datasets and added to them to ensure they covered a wide range and were representative. To improve the quality and variety of the dataset, we used methods like normalization and data augmentation. This allowed us to train and evaluate the models better.

When we dug deep into our experiment results, we found some interesting things. The CNN models were good at spotting and sorting out flaws in semiconductor pictures. They got it right over 90% of the time, no matter which model we used. This shows that deep learning works great for checking semiconductors. These models were able to notice tiny details and patterns in the semiconductor images. What's more, we found that the CNN models could handle different environments and lighting pretty well, which is important in semiconductor factories.

Despite these strong points, we ran into some limits and areas for improvement. One big issue was how CNN models tended to overfit when trained on small or uneven datasets. Also, training deep neural networks on big semiconductor datasets was complex. This created logistical challenges, making us think about how to use our resources well and scale our work.

When compared to the old ways of checking semiconductors, CNN models showed big improvements. They did better than the usual rule-based systems or machine learning tools, proving to be more accurate, faster, and better at handling distinct types of semiconductor images. By learning on their own from raw data, CNN models showed they were good at spotting small defects and sorting semiconductor parts every time.

Even so, putting CNN models to work in real-world factories comes with its own problems. These include the need for a great quantity of labelled data, powerful computers, and ways to understand what the models predict. Despite these roadblocks, the promise shown by CNN models highlights their ability to cause a revolution in how we check semiconductors. This opens the doors for better quality control and ways to reduce defects.

To wrap up, our study shows how CNN models have an enormous impact on semiconductor inspection. By using the good parts of deep learning methods and fixing the problems we found, makers can find new ways to make semiconductor products better and more dependable. Going forward, we need to keep up the research and development to overcome the current problems and make models using deep learning work even better for everyday semiconductor inspection in factories.

5.13 CONCLUSION

In conclusion, our study takes a close look at how deep learning CNNs can be used to inspect semiconductors. We showed that CNN models are accurate at finding and sorting out flaws in semiconductor images through careful testing and review. Our results prove that CNNs work well, with over 90% accuracy across different setups and samples. These findings show that deep learning methods have a huge effect on semiconductor inspection, giving makers new tools to control quality and reduce defects. We can't stress enough how valuable deep learning is in semiconductor testing. By using CNN models, makers can improve how they find defects, boost product quality, and cut production costs. Using deep learning methods marks a tremendous change in how we control semiconductor quality, allowing for automated, efficient, and exact defect spotting on a large scale. Looking forward to future studies should work on making CNN structures better, improving training methods, and finding new ways to use deep learning in semiconductor making. Also, working to create standard datasets, testing frameworks, and tools to understand the process will be key to helping more people use deep learning in semiconductor quality control and ensure it fits into industry practices.

REFERENCES

1. S. Zhou, Y. Chen, D. Zhang, J. Xie, and Y. Zhou, "Classification of surface defects on steel sheet using convolutional neural networks," *Mater. Technol.*, vol. 51, no. 1, pp. 123–131, 2017. https://doi.org/10.17222/mit.2015.335.
2. Y. Lecun, B. Yoshua, and H. Geoffrey, "Deep learning," *Nature*, vol. 521, no. 7553, pp. 436–444, 2015.
3. J. O'Leary, K. Sawlani, and A. Mesbah, "Deep learning for classification of the chemical composition of particle defects on semiconductor wafers," *IEEE Trans. Semicond. Manuf.*, vol. 33, no. 1, pp. 72–85, 2020. https://doi.org/10.1109/tsm.2019.2963656.
4. O. Russakovsky, J. Deng, H. Su, J. Krause, S. Satheesh, S. Ma, Z. Huang, A. Karpathy, A. Khosla, M. Bernstein, A. C. Berg, and L. Fei-Fei, "ImageNet large scale visual recognition challenge," *CoRR.*, 2014. http://arxiv.org/abs/1409.0575.
5. T. Wiatowski, and H. Boelcskei, "A mathematical theory of deep convolutional neural networks for feature extraction," *IEEE Trans. Inf. Theory*, vol. 64, pp. 1845–1866, 2018. https://doi.org/10.1109/TIT.2017.2785642.
6. S. Ren, K. He, R. Girshick, and J. Sun, "Faster R-CNN: Towards real-time object detection with region proposal network," in *Proc. Neural Information Processing Systems*, Montreal, QC, Canada, 7–12 December 2015.
7. V. Gyanendra, and D. Anamika, "A Handheld Gun Detection using Faster R-CNN Deep Learning," in *Proc. 7th Int. Conf. Computer and Communication Technology*, Allahabad, India, pp. 84–88, 24–26 November 2017.
8. K. Imoto, T. Nakai, T. Ike, K. Haruki, and Y. Sato, "A CNN-based transfer learning method for defect classification in semiconductor manufacturing," *IEEE Trans. Semicond. Manuf.*, vol. 32, no. 4, pp. 455–459, November 2019.

6 Machine Learning for Semiconductor Devices

Nivedita Manohar Mathkunti and Manjulata Badi

6.1 INTRODUCTION

The expansion and broad implementation of the new technology of semiconductors has been the main steering force behind the 21st-century computer revolution. Furthermore, the fabrication of semiconductors is now recognized to be a major focus on public innovation in various countries. For instance, the European Chips Demonstration Act (EU), passed in 2023, and the CHIPS and Science Act (US), passed in 2022, have both enacted significant subsidizing measures to promote the production of semiconductor chips, communication, engagement, and employment. Elite performance semiconductors are becoming increasingly popular as a result of emerging applications like computer-based artificial intelligence (CAI) [1] that demand large amounts of processing power.

For this reason, a number of approaches have been developed and proposed, including 3D bundling, silicon photonics, and heterogeneous combination, and these are gateway to overall semiconductors [2]. The unifying aim of all approaches is to attain the smallest line width (pitch) at the expense of exceedingly complex structures. These developments allow chips or parts of chips to operate independently by using connections at the wafer and chip levels [3]. When compared to conventional bundling schemes, they could create a very confusing framework. The complexity of the strategy and its creation increases the production of successful framework.

To collect data and evaluate cycle boundaries, item usefulness, layer thickness, arrangement position, material piece, and electrical qualities metrology use a variety of estimate methodologies, changing units of estimate and equipment [4]. According to Abd Al Rahman and Mousavi [5], the market for semiconductor review and estimation hardware's includes a variety of devices such as cover estimation hardware, film thickness, overlay accuracy, three-layered form, example and non-design wafer imperfection examination devices, and veil investigation devices. Process control is metrology's primary responsibility in the semiconductor industry, and metrology is essential to the industry's continued viability. For instance, 3D NAND devices use materials such as thick, hard coverings; extremely loaded designs; and more than 100-layer matches. Elevated geological heterogeneity, ambiguity in the material, and wafer stress are some of the factors that enhance the difficulty of achieving accurate estimations [6]. Additionally, the production of dynamic irregular access memory (Measure) further complicates the process.

DOI: 10.1201/9781003508304-6

However, there will always be a need for technological devices with greater functionality. Future semiconductor techniques, such as advanced bundling, are expected to be denser and require a lot more intricate architecture. As devices become smaller and more three dimensional (3D) in design, metrology becomes increasingly necessary. Over half of the manufacturing stages for some commodities involve estimation or depiction. The semiconductor industry is heading toward a future when it should be possible to determine the nature and condition of every particle inside a 3D device. The base interconnection pitch is now submicron, and in some applications, the allowable layer thickness for semiconductor bundling has expanded by more than 100 times in the past ten years. The pattern toward expanding framework thickness is expected to go on from here on out, presenting huge difficulties in both planning and actual handling. Several investigations have shown that analyzing the dependability of upcoming semiconductor procedures is the main deterrent [7]. Many semiconductor devices have multiple components that have varieties of functionalities. As a result, more research and fortifications are needed to support current metrology advancements [8].

During semiconductor fabrication operations, completing the goal of a thorough inspection of partially produced products necessitates a limited quantity of metrology equipment and time. The company now undertakes haphazard inspection and testing for quality control to save money, but this doesn't always result in thorough quality control or guarantee exceptional results. The metrology procedures that are now in use need examination during particular dealing stages, and their estimates are fulfilled at clearly identified metrology stations, resulting in pointless handling time. Virtual metrology (VM) may therefore address the consistent quality of testing and review for advanced semiconductor products [9]. To evaluate the outcome of the discussed fab interaction, VM makes use of cycle factors, creation apparatus information, and tested wafers for genuine metrology information. Although all products cannot be subjected to actual metrology, VM may be used to meet the need for a "complete" inspection. This method can replace the inline continuous inquiry with detached review given the proper setup and testing of virtual machine models [10]. VM, also referred to as delicate detecting, has found useful uses in different assembly industries in the past. Though these applications really need further research, they have benefited greatly from the semiconductor industry's successful use of virtual machines. For instance, to lower manufacturing costs, factories aim to lower sample rates. The intelligent sampling decision (ISD) system supports VM, however, its set rate limits adaptability. This study proposes an automated sampling decision (ASD) method that dynamically adjusts sampling rates in real time based on VM correctness to guarantee efficiency and long-term performance. This semiconductor process was developed by Cheng and others [11] to accomplish the objective of complete mechanization in the auto-wheel creation business. The inventors demonstrated a workable arrangement of three machines, one inline, one drill, and one additional unconnected metrology station, that work together to create an automated wheel production facility. The usefulness of this VM in the making of copper-clad overlay (CCL), a basic component for on-paper circuit sheets used in a variety of devices, was investigated by Kim and others [12]. Using verified data from a CCL manufacturer, three quality parameters were the subject of expectation during the therapy. Similar to how carbon fiber creation was integrated into the AVM framework, makers were able to track creation information, initiate a creation information traceback (PDT), link it to a specifically manufactured workpiece, and complete a thorough assessment of the carbon fiber creation process. The application of virtual maintenance technology to the commercial motors in aviation was demonstrated well by Enrico and others [13], and the applicability of VM for ultra-accurate device production and machining was discovered in both commercial motors in aviation and ultra-accurate device production. Therefore, by examining these continuing patterns of assembly, it can be safely concluded that VM is a crucial mechanical advancement of Industry 4.0 that goes beyond traditional metrology target expectations. It is imperative that ventures comprehend that virtual machine foundations now include a substantial portion of Industry 4.0 logic such as internet of things (IoT), digital physical frameworks, modern massive information, and zero-imperfection assembling [14].

Future semiconductor assessments are expected to heavily rely on VM. As bundling complexity increases, inclusion size decreases, and advanced bundling techniques such as wafer-to-wafer and pass-on-to-wafer are dynamically replacing normal cycles, necessitating more ongoing research. Additionally, VM can play a big role in the design of cutting-edge bundling, which is recognized by several assessments as a major challenge as construction thickness increases. The advent of quicker artificial intelligence and the convenience of open source AI-ML (artificial intelligence–machine learning) bundles has led to the recent blossoming of semiconductor APC (advanced phase control) speculations, including virtual machines (VMs), which are helpful for development bundling. To overcome prophetic exactness issues posed by sophisticated innovative computations, constant innovation in VMs is required [15].

A thorough production and the potential implications of VMs for this sector are vital. For this reason, this chapter undertakes such a leading effort. Appropriate VM topics related to semiconductor fabrication are compiled and analyzed, and the best available research is summarized in this chapter. The chapter audits and examines the adopted virtual machine tactics in every cycle bunch in light of the semiconductor manufacturing process hierarchy. Analogously important, fundamental analysis is conducted to investigate persistent challenges in the further implementation of semiconductor fabrication as well as a comparative analysis of potential futures.

6.2 ADVANCEMENT OF SEMICONDUCTOR GADGETS

Within the hardware realm, the operation of semiconductor devices continues to be a showcase for human creativity and technological advancement. Over the years, semiconductor devices have seen notable advancements and remarkable improvements in their direction. These devices, which started off as diodes and semiconductors, have evolved into the cornerstone of modern invention and are responsible for the sophisticated era we live in today. This section begins with an examination of the verifiable trajectory of semiconductor devices, highlighting their noteworthy accomplishments and the dynamic changes that have shaped their progress. By means of this journey, we have a deeper understanding of the intricately woven artwork of logical discoveries and design achievements that have propelled semiconductor devices into their current critical condition.

6.2.1 Definition: The Semiconductor Gadgets

Semiconductor devices are electronic components that are based on a semiconductor's unique electrical properties. These substances lie midway between the guide and separator domains in terms of conductivity. Important semiconductor materials include natural semiconductors such as silicon, germanium, and gallium arsenide. These electronic components, which make use of the remarkable electrical capabilities of semiconductors, are the foundation of many creative applications.

6.2.2 The Advancement in the Evolution of Semiconductor Gadgets

This section provides a clear and concise synopsis of the evolution of semiconductor innovation. It covers the pivotal moments, advancements, and trends that have shaped this area, from the discovery of early semiconductors to the present-day coordinated circuits. This section provides a foundational understanding of the empirical environment, paving the way for understanding the current state of affairs and potential future directions for semiconductor devices. In particular, the persistent scaling of complementary metal oxide semiconductor (CMOS) innovation, per Moore's law, has long been the primary driver in the semiconductor industry. The continuous reduction in the size of semiconductors and circuits has enabled significant advancement in computing power and the proliferation of electronic devices across various applications.

6.2.2.1 Early Semiconductor

The term "early semiconductor" refers to the fundamental configuration of semiconductors created in the latter half of the 1940s and the middle of the 1950s. We have discovered another impact, now referred to as the semiconductor impact, in our latest investigations using point-contact rectifiers. This effect is dependent upon the use of a small portion of semiconductor material as opposed to the conventional approach of using the majority of such materials. These were the primary strong-state components used in many practical applications, taking the place of vacuum tubes. When compared to vacuum tubes, these early semiconductors were smaller, more reliable, and consumed less power, which was a significant advancement in hardware. The investigations and preliminary work conducted during this period established the framework for the modern semiconductor.

6.2.2.2 Integrated Circuit

Kilby invented the coordinated circuit, which fundamentally altered the hardware landscape by combining several components on a single semiconductor substrate. Kilby's seminal work in the latter half of the 1950s laid the foundation for the advanced era of high-performance, reduced-size electronic devices. A smaller version of an electrical circuit, consisting of resistors, capacitors, diodes, and semiconductors is called an integrated circuit (IC). These are produced mostly on silicon, a tiny semiconductor substrate. These elements work together to accomplish certain tasks such as handling, memory storage, and improving signals. When many components are combined into one chip, it is possible to reduce size, enhance performance, and use less energy than when separate electronic circuits are used. Coordinated circuits are crucial for powering several modern devices.

6.3 CURRENT INNOVATIVE PATTERNS

6.3.1 Arising Advances

Several state-of-the-art developments have been made in the last century and a half. Innovative assembly, information, and video innovations are working together to create new advancements that are more effective. The development of fluid biopsy technology tackles a crucial phase in the ongoing battle against illness. Fluid biopsies offer several advantages over traditional tissue biopsies. They serve as a good alternative right away in cases where tissue biopsies make no sense. Furthermore, fluid biopsies provide a comprehensive view of the patient's overall health, whereas tissue biopsies only provide fragments of information based on the sample. In fact, circulating growth DNA is frequently detected in fluid biopsies because it frequently migrates from cancer tissue into the bloodstream. With the advancement of convolutional brain structures and deep learning, machine vision has begun to surpass human capabilities in image recognition. Currently, machine vision innovation is showcasing a wide range of anticipated applications in several domains, including autonomous driving, medical diagnosis, protection guarantee assessment, water level monitoring, gardening, and more. Convolutional neural networks (CNNs), which can be trained using the backpropagation algorithm, consistently outperform other deep learning structures in tasks like image and speech recognition. They also need fewer boundary evaluations than other feedforward deep learning models, which make them an attractive option in the great learning space. The potential of quantum computers is infinite, but their development is extremely difficult and expensive.

The reason why the processing power of small quantum personal computers has not surpassed that of supercomputers is easily understood. However, IBM advanced the use of quantum computing in 2016 by being the first company to provide cloud computing services for quantum computers to the general public. More than 20 scientific publications that are awaiting dissemination now have an exploratory platform thanks to the invention. Currently, more than 50 organizations worldwide, including large initiatives and countless new businesses, are working to bring quantum PCs to reality. People believe that the quantum era has come as a result of these developments.

6.3.2 The Advantage of the Rising Advancements

Innovation and science play crucial roles in advancing human progress. No matter how much technology advances, it is important to remember that machines are still products of human invention, meant to be used by humans. Those who benefit from these mechanical devices shouldn't suppress their innate desire to labor, as an excessive reliance on machines might lead to social disapproval. However, this does not imply that new developments obstruct human progress. Similar to other aspects of daily life, innovation is a two-sided affair. Our choices determine how we use it. It is absurd to disparage the contribution that research and innovation may make to human progress since individuals consistently demonstrate their prevailing knowledge.

6.3.3 Execution Improvement

Shrewd chips are the supporting innovation needed for 5G and the IoT, which will promote these technologies further and improve smart devices, smart homes, and other fields, in addition to improving information inclusion and digitalization. The semiconductor sector will bring about more notable advancements and beneficial opportunities, and demand for chips will only grow. The growing consciousness and deep learning of humans mean that the semiconductor industry will need to meet increasing performance requirements in terms of energy efficiency and computation. As of right now, the semiconductor business will see considerable development in the areas of PC engineering innovation and computation streamlining. In contrast to earlier generations, modern electronics are smaller and more versatile, offering a higher degree of customization. To satisfy the modest needs of customers, the semiconductor industry will primarily support new radio-frequency devices and smaller chips in this particular situation.

6.4 SEMICONDUCTORS IN AI/ML WITH DIFFERENT METHODOLOGY

The Semiconductor Equipment and Materials International's (SEMI) Equipment Communications Standard/Generic Equipment Model (SECS/GEM) protocol makes it easier for host computers and semiconductor equipment to communicate in the semiconductor manufacturing sector. SEMI developed SECS/GEM with the goal of enhancing factory automation through standardized communication. In smart manufacturing systems, hosts that manage and gather data from the equipment make up the edge layer, while the equipment itself is contained in the equipment layer. However, the existing SECS/GEM interface lacks dynamic data analysis at the hardware level, which results in latency issues when transferring input to the edge/cloud layers for analysis.

To address these latency issues and enable dynamic data processing at the equipment level, the research work [16] proposes a novel communication protocol for the SECS/GEM interface. The present SECS/GEM interface adds significant latency when sending data from equipment to edge/cloud layers because it does not offer dynamic data analysis.

Nguyen [16] conducted experiments utilizing SECS/GEM messages to quantify the lag of data variable transfers from the hardware to the host. A notable lag that could affect productivity was revealed by the data. To solve this, they proposed an architecture for the system that includes an onboard "Data Analysis Engine" module. This module makes it possible for equipment to perform host-specified analysis tasks, hence eliminating the need for data transfers to the host.

The authors included a new "Stream 22" in the SECS/GEM protocol, which allows the host to build, link, and start custom and predefined analytic operations on the equipment as well as report and identify outcomes. By enabling the host to assign data analysis tasks to the equipment dynamically and collect the results, this proposed protocol can reduce network bandwidth usage and improve semiconductor smart manufacturing efficiency by enabling real-time reactions [16].

The defect detection model is built using You Only Look Once version 2 (YOLOv2) as per the reference model. YOLOv2 is a well-known object detection algorithm that performs detection through regression, using many different pooling layers with convolution to transform the input image into a 3D tensor. The input images to the YOLOv2 model are of size 448 × 448 pixels, which improves the resolution and fine-tunes the classifier for better performance. To improve the critical dimension scanning electron microscope's (CD-SEM) gray images by its features, the contrast and brightness of the images are dynamically adjusted. This is achieved using image linear blending technology. Two images, the original image I_0 and a second transparent channel image I_1, are fused linearly with respect to weights. The blending formula is:

$$g(x) = \lambda I_0(x) + (1 - \lambda)I_1(x) \tag{6.1}$$

In the blended image g, the resulting pixel value at position x is considered where λ is the weight controlling the fusion of the pixel values at corresponding positions in the two images [17].

The system processes CD-SEM gray images automatically by extracting them from the CD-SEM server database, preprocessing them, and then using the defect detection model to identify defects. If a defect is detected, the system saves this defected output into CD-SEM data as a review image. It also triggers an alarm email to notify engineers. Engineers can view detailed CD-SEM data and review images through a custom user interface. The system supports batch detection of CD-SEM images to improve efficiency. For example, it can process up to 188 gray images in a batch, displaying the results with bounding boxes indicating defects (red for defects, green for no defects). The system reduces the need for engineers to manually review every image and enhances the inline defect inspection capability, thus improving the economic benefits for the factory. By noticing faulty wafers in real time and notifying engineers promptly, the system helps reduce the impact on products and improves yield [18, 19].

In this study, the researcher Nguyen, H. developed a robust supervised deep learning training scheme aimed at accurately classifying and localizing various defect types in SEM images with high precision. The proposed ensemble model integrates ResNet101, ResNet50, and ResNet152 designs as spines and utilizes experimentally selected models based on preferences strategically ensembled to get out after combining the predictions from the models. This approach enhances performance in both the classification and detection of defects.

The model efficiently categorizes faulty classes, such as line collapses, micro-bridges, and micro-gaps and handles the changing grades of pixel-level faulty scenarios within these categories. Additionally, it precisely regresses the regions of the faults by providing bounding boxes signified by width, center coordinates, height, and confidence score of detection.

To further optimize fault detection, unsupervised machine learning strategy is applied to get noiseless images of SEM without any clear ground truth. This method benefits in the removal of false positive faults and minimizes the impact of stochastic noise on pixels that are structured. The analysis of power spectral density shows that only the noise of high-frequency components is exaggerated. Along with preserving the frequency of low components related to the sound structure of device features.

This defect inspection pipeline enhances the accuracy of defect detection. This process will not change any dimensions of wire length/wire space (L/S) in aggressive pitches. As circuit designs continue to shrink to comply with Moore's law, these conventional defect inspection methods become less effective, often leading to false defect detections and erroneous metrology. A model based on deep learning addresses these limitations, improving the classification and detection localization of various defect categories with greater precision and accuracy [20].

As semiconductor manufacturing moves toward the 3 nm node, reducing edge placement error (EPE) is crucial for proper IC device functioning. EPE evaluates the reliability of produced patterns in multi-patterning processes by integrating overlaying and critical dimension (CD) defects. Recent ML advances have provided new opportunities to improve the performance and efficiency of EPE optimization procedures.

Kyon and others [21] survey recent research applying machine learning/deep learning to enhance virtual overlay metrology, reduce overlay error, and improve mask optimization methods for EPE reduction. The review discusses objectives, datasets, input features, models, key findings, and limitations. Results show ML's great potential in improving EPE in semiconductor manufacturing.

Samsung started the "initial" 3 nm production in 2022, using gate-all-around architecture, offering 16% higher transistor density, 23% higher performance, and 45% lower power compared to an unspecified "5 nm" process. The Taiwan Semiconductor Manufacturing Company (TSMC) announced volume production of its "3 nm" N3 process in December 2022, with the refined N3E process starting in H2 2023. It is the semiconductor fabrication process. At IEDM (International Electron Devices Meeting) 2022, TSMC disclosed that N3 has a 45 nm contacted gate pitch and N3E a 23 nm minimum metal pitch, with SRAM cell areas of 0.0199 μm^2 for N3 and 0.021 μm^2 for N3E. As nodes shrink, more intellectual is required in filling, as it can affect timing and signal integrity of all layers' fillings [21].

The paper, Chiplet Heterogeneous-Integration AI processor, discusses the design of a chiplet-based AI processor named ABSx, which integrates multiple neural processing units (NPU) and high-bandwidth memory (HBM3) chiplets on a multilayer redistribution layer (RDL) interposer. Key design aspects include high-speed data channels, with an emphasis on the necessity for high-speed data channels between NPUs and HBM to facilitate parallel processing and near-memory processing, which is crucial for AI tasks.

Interposer Design: Considerations for signal integrity, power distribution, and high-density wiring to ensure effective communication and power management across the chiplets.

Intra-Chiplet NPU Architecture: Design of NPU architecture to sustain continuous data feeding into the floating-point units, ensuring efficient processing.

Bonding Reliability: Addressing static and dynamic bonding reliability issues caused by the numerous micro-bumps, essential for maintaining the structural integrity of the processor.

Thermal Stability: Ensuring thermal stability by analyzing heat flux and mutual heat influence and implementing effective heat removal strategies to manage the thermal load.

Inter-Chiplet Link: Highlighting the importance of the inter-chiplet link and its impact on the overall performance efficiency of the processor.

The design criteria focus on optimizing performance efficiency, ensuring reliable chiplet integration, and maintaining thermal stability. The chiplet architecture is used to address the difficulties that have arisen as a result of the data explosion in computing, which is necessary for large-scale AI models. The AI-based chiplet ABSx is a suitable solution for AI's data processing requirements, providing a way ahead for handling the complex workload in line with maintaining high-performance dependability [22].

In Yoo and others [23], the researchers present the working of natural language processing (NLP) techniques to evaluate text data related to apparatus maintenance as well as quality concerns in the semiconductor manufacturing industry. In the view of assembling insightful information from this unstructured data, the researchers undertook a number of crucial steps in the development of an internal NLP engine. In the beginning, the researchers preprocessed all text data, which included details on equipment maintenance as well as quality issues in sequence to create a corpus. A semiconductor dictionary was used to do analysis of quality troubles on the semiconductors. These also use the bag-of-words, bag-of-*N*-grams techniques, and term frequency-inverse document frequency (TF-IDF) to quantify the document and extract important features. The visualization of phrases and the association of key phrases in different factories are analyzed, and significant differences between equipment types were found with the use of chi-square tests. In the end, burst analysis identified the times when the frequency of significant phrases grew greatly.

The findings showed that different factories and models had different quality issue trends and equipment maintenance characteristics. This offered insightful information that could help semiconductor manufacturers increase productivity and quality.

The authors' use of NLP techniques to analyze unstructured text data demonstrates the potential of this approach to extract meaningful insights from the wealth of information available in semiconductor operations. By uncovering patterns and differences in quality and maintenance issues, the NLP analysis can help manufacturers optimize their processes and enhance overall performance.

Yoo and his team present a comprehensive methodology that combines technology computer-aided design (TCAD) simulations with advanced ML techniques to achieve multi-target optimization for BCD (Bipolar-CMOS-DMOS (Double – diffused metal oxide semiconductor)) processes. The key aspects of the approach are data generation, where it uses 150 Latin hypercube sampling points to thoroughly cover the parameter space and generate essential data for model training through TCAD simulations. This foundational data is used to establish accurate relationships between process variables and the electrical characteristics (ET-spec) of target devices. The next step is regression modeling, and models are trained to forecast the ET-spec of equipment based on procedure parameters. An auto-regression model is selected for its high fitness, allowing the use of surrogate models instead of direct TCAD simulations to significantly reduce computational costs during iterative optimization processes. In the optimization process, a genetic algorithm (GA) is employed to efficiently search for the global minimum by exploring a broad solution space. The optimization process uses a weighted sum of fitness scores to manage the multi-target nature of the problem, ensuring that different device parameters are optimized simultaneously.

To examine different parts of the Pareto front, multiple GA solutions are started from distinct beginning points. TCAD simulations are used in validation to verify the efficacy of the optimized process parameters that the GA found. This stage guarantees that the suggested fixes are workable from a practical standpoint as well as conceptually sound.

The new methodology is compared with existing optimization flows in the comparison and benefits step. Existing optimization flows target devices sequentially and necessitate numerous iterations. The proposed strategy, on the other hand, achieves an 87% targeting rate and drastically cuts the turnaround time from 30 days to just three days by optimizing all variables across all devices at once.

This methodology efficiently solves the difficulties and trade-offs inherent in BCD process development, offering a reliable solution for next-generation semiconductor production by merging AI-based regression models with TCAD simulations and using a GA for optimization [24].

Through an examination of the neural network physically unclonable function (NN PUF), implemented on the SAKURA-G field-programmable gate array (FPGA) evaluation board, the article performance evaluation of AI authentication device implemented on SAKURA-G addresses security problems related to AI. The NN PUF takes advantage of the variety in large-scale integration (LSI) production to produce unique identifications (IDs) for device verification by combining a neural network and the physically unclonable function. The paper [25] evaluates the NN PUF's performance based on diffuseness, steadiness, randomness, and uniqueness. The methodology includes NN PUF implementation, PUF circuit, and authentication methods.

The NN PUF integrates a multilayer neural network using a look-up table (LUT) network framework. A neural network has an input level of 1,024 LUTs. Two layers are buried using 360 as well as 60 LUTs, respectively. A production layer includes ten LUTs. The NN is intended to identify numbers written by hand using the MNIST dataset, with 784 bit inputs as a PUF constraint.

For the physically unclonable function circuit-authentication mode, the NN PUF uses challenge-and-response pairs for device authentication. The end product of the neural network is attached to a circuit – for example, a flip-flop of type D that generates PUF responses by extracting differences in signal propagation delays. This variance in delays, due to LSI manufacturing differences, helps generate unique IDs.

The evaluation metrics are randomness, steadiness, diffuseness, and uniqueness. Randomness measures the uniformity of 0/1 occurrences in the produced ID, which is evaluated by the Hamming weight, being marked to part of the ID size. Steadiness assesses the duplicability of the ID for the same input challenge, estimated with the same challenge intra-Hamming distance (SC Intra-HD). Diffuseness determines how the ID changes with different input challenges, assessed by dissimilar challenge intra-HD (DC Intra-HD), and uniqueness evaluates the ID variance between different devices, assessed by the same challenge inter-HD (SC Inter-HD).

The methodology provided a comprehensive framework for evaluating NN PUF performance on advanced FPGA boards, emphasizing the importance of process variations in PUF performance [25, 26].

A methodology for effectively employing deep learning techniques to search for similarity in wafer defect maps is presented in the paper "Wafer defect map similarity search using deep learning in semiconductor manufacturing." The first step in the procedure is to convert the raw defect scan images into grayscale representations of defect density, noise reduction, and defect distribution.

For feature extraction, a pretrained CNN, namely ResNet, is used. After being trained on a sizable dataset, this network takes the grayscale images and extracts high-dimensional feature vectors that capture important defect properties.

A model of *K*-nearest neighbor search is developed with the use of feature space, where vectors are generated. The top *n* images are identified, which are the same as the query images, with respect to their feature vectors, and these enable the effective extraction of similar defect maps from the bigger datasets.

While receiving the search model remotely, the solution makes use of the gRPC framework to simplify communication between the client and server so that clients can conduct inquiries locally. The system's exceptional accuracy and memory rates were demonstrated by the top three searches, which had 100% accuracy and a 92% recall rate. That the search feature finds the top 10,000 similar photos in less than four seconds is evidence of how fast it works. This method significantly increases the accuracy and efficiency of defect analysis in semiconductor production [27].

The paper [28] presents a novel approach to accomplish numerical procedure handling in high-capacity semiconductor device manufacturing, specifically focusing on the detection and prediction of film breadth in chemical vapor deposition (CVD) processes.

The key aspects of the proposed methodology include VM implementations, which leverages production parameters to predict measurement values in real time, reducing the need for physical sampling and compressing the cycle time. A limitation of most of the current VM strategies used in universal nonlinear models is that they struggle to capture sudden changes in the fluctuating production process. Additionally, traditional ML methods only provide predicted values without assessing the confidence of the results, leading to lower accuracy when dealing with high-dimensional data and small sample sizes. The lasso-Gaussian process regression (LGPR) is a way to address these limitations; the [%%] LGPR model is grounded on the just-in-time learning framework. LGPR combines the strengths of lasso regression for feature selection and Gaussian process regression for accurate prediction and uncertainty quantification. The experimental results demonstrate that the LGPR-based VM system is more stable and accurate compared to traditional approaches, enabling better real-time prediction of film thickness in CVD processes. This, in turn, supports effective statistical process control and early detection of abnormal conditions in high-volume semiconductor manufacturing [28–30].

The paper [29] focuses on defect analysis in semiconductor manufacturing, organized keenly into three phases: defect classification, defect trend monitoring, and detailed classification.

A CNN-based transfer learning method is used for automatic defect classification to support engineers in their analytical work. The methodology involves using deep learning technologies to assist engineers in the first and third phases of defect analysis, reducing labor costs by one-third compared to manual inspection work. Transfer learning is introduced to reduce the amount of labeled training data required for classification, especially when reliable labeled data is limited, thus improving classification accuracy using a limited number of reliable labeled data.

The paper [31] discusses the adoption of inaccurate supervision due to inconsistent manual labeling and the utilization of transfer learning to reduce the need for a large amount of training data with ground-truth labels [31, 32].

6.5 DISCUSSION

Seeing the global demand for semiconductors for ICs, it is greatly necessary to promote IC manufacturing with the use of AI and ML technology. A Deloitte white paper that is exploring the opportunities available to IC companies [31] says that the

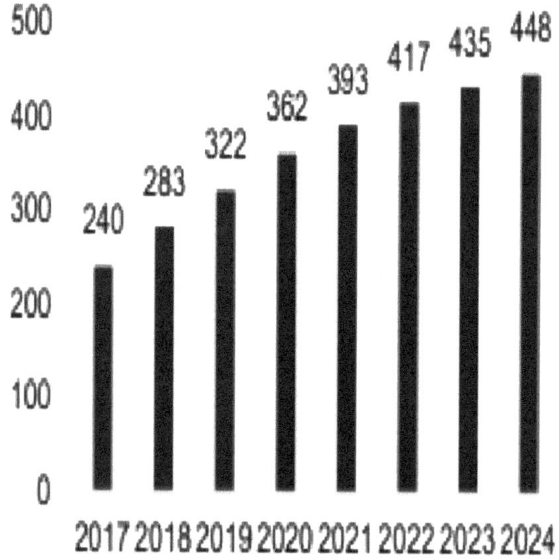

FIGURE 6.1 The global revenues of electronic consumers.

Source: ESDM Industry, KPMG in India.

economy of the industry had previously doubled and now tripled. The huge demand for ICs triggers AI chip development; this involves almost all ML algorithms and deep learning, which falls under the AI umbrella. At this moment, schemes for the growth of semiconductor manufacturing in India should be announced. The growth of the electronic sector, for which semiconductors are the building blocks, is discussed in [34]. The global revenues of electronic consumers are depicted in Figure 6.1.

The world's best economies have drastically changed due to the rise in IC manufacturing. The point to be noted here is that there has been a 360-degree evolution not just in AI or semiconductors but in both.

The world revenue of consumer electronics shown in Figure 6.1 [34] gives us an idea of the drastic hike in the demand for electronics, which have semiconductors as their basic building blocks. The gradual development in the demand for ICs involves many technologies such as IoT, AI/ML, and deep learning. The global demand for semiconductors from 2020 to 2030 is shown in Figure 6.2.

This chapter gives us a glimpse of the applications of all ML algorithms in semiconductor manufacturing and defect detection. The study is about machine learning for semiconductors. It reveals that with the help of deep learning algorithms, fault detection of SECS/GEM can be done. In the design of AI-based processor, semiconductors

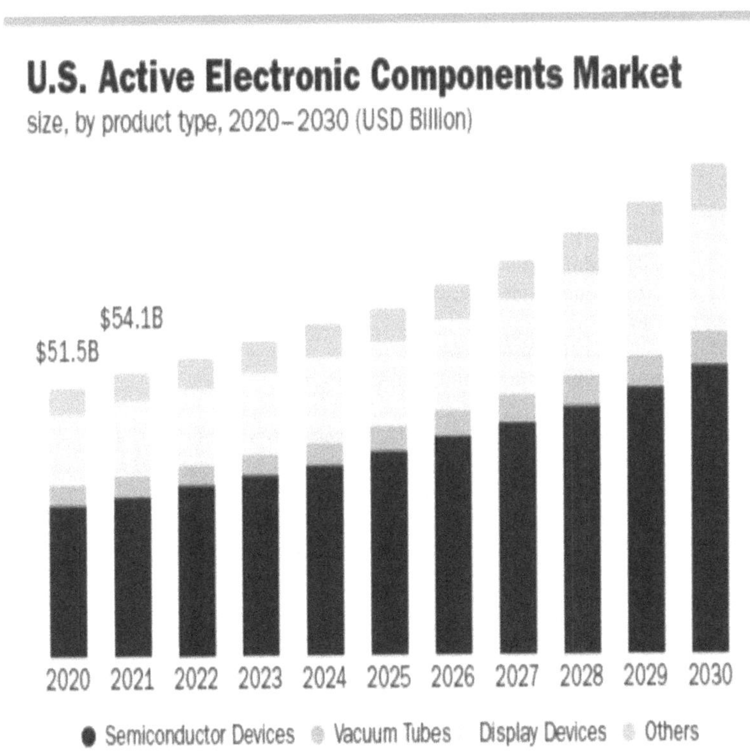

FIGURE 6.2 Demand for semiconductors from 2020 to 2030 in USA.

can also be manufactured. NLP is used to extract features. Classification of the semiconductor can also be performed through machine learning algorithms.

Figure 6.2 highlights the drastic growth in the production of semiconductors, which are in turn used in the production of components like diodes, transistors, ICs, vacuum tubes, display devices, and optoelectronics. These components end up in various sectors like automotive, manufacturing, aerospace and defense, health care, and networking and telecommunication. Semiconductor production is concentrated in a few regions, with 54.5% of the revenue share being in Europe, Asia-Pacific, and South Africa. Toshiba Corporation, Microchip Technology, NXP Semiconductors N.V., Texas Instruments, and Qualcomm are the major contributors to this growth.

6.6 CONCLUSION

In this study, many different AI-based algorithms are applied to enhance the production, and detection of faults for semiconductors. From the present study, as well as by observing the growth in the market, it can be concluded that there are many opportunities for semiconductor-based IC design as well as IC production. To meet this demand, many ML algorithms and deep learning algorithms can be applied to enhance production and reduce defects, and more research is also necessary to further digitalize this world.

REFERENCES

1. Leng, S., R. S. Tan, K. T. C. Chai, C. Wang, D. Ghista, and L. Zhong. "The electronic stethoscope." *Biomedical Engineering Online* 14 (2015): 1–37.
2. Bogaerts, W., and A. Rahim. "Programmable photonics: An opportunity for an accessible large-volume PIC ecosystem." *IEEE Journal of Selected Topics in Quantum Electronics* 26, no. 5 (2020): 1–17.
3. Aubusson, Russel. "Wafer-scale integration of semiconductor memory." PhD diss., Middlesex Polytechnic, 2011.
4. Lapkina, I., and M. Malaksiano. "Estimation of fluctuations in the performance indicators of equipment that operates under conditions of unstable loading." Восточно-Европейский журнал передовых технологий 1, no. 3 (2018): 22–29.
5. Abd Al Rahman, M., and A. Mousavi. "A review and analysis of automatic optical inspection and quality monitoring methods in electronics industry." *IEEE Access* 8 (2020): 183192–183271.
6. Capra, M., B. Bussolino, A. Marchisio, G. Masera, M. Martina, and M. Shafique. "Hardware and software optimizations for accelerating deep neural networks: Survey of current trends, challenges, and the road ahead." *IEEE Access* 8 (2020): 225134–225180.
7. García de Arquer, F. P., D. V. Talapin, V. I. Klimov, Y. Arakawa, M. Bayer, and E. H. Sargent. "Semiconductor quantum dots: Technological progress and future challenges." *Science* 373, no. 6555 (2021): eaaz8541.
8. Guha, S., H.-S. P Wong., J. A. C. Incorvia., and S. Chowdhury. *Future Directions Workshop: Materials, Processes, and R&D Challenges in Microelectronics* (2022). https://doi.org/10.21236/ad1188476
9. Dreyfus, P-A., F. Psarommatis, G. May, and D. Kiritsis. "Virtual metrology as an approach for product quality estimation in Industry 4.0: A systematic review and integrative conceptual framework." *International Journal of Production Research* 60, no. 2 (2022): 742–765.
10. Grooters, T. "Analysis of automated virtual machine generation and automation around system testing at TKH airport solutions." Master's thesis, University of Twente, 2023.

11. Cheng, F-T., Y-S. Hsieh, C-F Chen, and J-R. Lyu. "Automated sampling decision scheme for the AVM system." *International Journal of Production Research* 54, no. 21 (2016): 6351–6366.
12. Kim, M., S. Kang, J. Lee, H. Cho, S. Cho, and J. S. Park. "Virtual metrology for copper-clad laminate manufacturing." *Computers & Industrial Engineering* 109 (2017): 280–287.
13. Enrico, Z. I. O., F. A. N. Mengfei, Z. E. N. G. Zhiguo, and K. A. N. G. Rui. "Application of reliability technologies in civil aviation: Lessons learnt and perspectives." *Chinese Journal of Aeronautics* 32, no. 1 (2019): 143–158.
14. Umasankar, M., S. Padmavathy, and N. Prakash. "Product quality through process improvement-a pathway to zero defects." *International Journal of Recent Technology and Engineering* 8, no. 4 (2019): 4286–4298.
15. Ruperao, P., P. Rangan, T. Shah, V. Thakur, S. Kalia, S. Mayes, and A. Rathore. "The progression in developing genomic resources for crop improvement." *Life* 13, no. 8 (2023): 1668.
16. Nguyen, H. H. "Based on deep learning CD-SEM image defect detection system." In *Dynamic AI Computation Tasks with SECS/GEM in Semiconductor Smart Manufacturing. 2022 International Symposium on Semiconductor Manufacturing (ISSM)*, 2022, pp. 1–4. https://doi.org/10.1109/cstic55103.2022.9856857.
17. https://www.cimetrix.com/blog/machine-learning-in-smart-manufacturing-technologies-you-should-be-tracking-now.
18. Kim, J., Y. Nam, M-C. Kang, K. Kim, J. Hong, and S. Lee. "Adversarial defect detection in semiconductor manufacturing process." *IEEE Transactions on Semiconductor Manufacturing*, 34, no. 3 (2021): 365–371.
19. Bappaditya, D., D. Goswami, S. Halder, K. Khalil, P. Leray, and M. Bayoumi. "Deep learning-based defect classification and detection in SEM images." (2022). https://doi.org/10.1117/12.2622550.
20. Ngo, A. T., B. Dey, S. Halder, S. De Gendt, and C. Wang. "Machine learning-based edge placement error analysis and optimization: A systematic review." *IEEE Transactions on Semiconductor Manufacturing*, 36, no. 1 (February 2023): 1–13. https://doi.org/10.1109/TSM.2022.3217326.
21. Kwon, Y., J. Han, Y. P. Cho, J. Kim, J. Chung, J. Choi, S. Park, I. Kim, H. kwon, J. Kim, H. Kim, W. Jeon, Y. Jeon, M. Cho, and M. Choi. Chiplet Heterogeneous-Integration AI Processor, "International conference on electronics, information, and communication (ICEIC)." (2023). IEEE. https://doi.org/10.1109/ICEIC57457.2023.10049867.
22. Kobayashi, D., S. Ito, S. Yasuda, and T. Iuchi. "Application of natural language processing in semiconductor manufacturing." In *2022 International Symposium on Semiconductor Manufacturing (ISSM)*, 2022, pp. 1–3. https://doi.org/10.1109/ISSM55802.2022.10026893[1].
23. Yoo, J., Y. Jeon, D. Jung, J. Kim, J. Ryu, U. H. Kwon, and D. S. Kim. "Machine-learning based TCAD optimization method for next generation BCD process development." In *2021 33rd International Symposium on Power Semiconductor Devices and ICs (ISPSD)*, 2021, pp. 279–282.
24. Takemoto, S., K. Shibagaki, Y. Nozaki, and M. Yoshikawa, "Small scale and low latency oriented neural network physical unclonable function and its evaluation." *IEEE Transactions on Electronics, Information and Systems*, 140, no. 12 (December 2020): 1297–1306.
25. Nozaki, Y., and M. Yoshikawa. "Performance evaluation of AI authentication device implemented on SAKURA-G." (2023) Unpublished.
26. Doe, J., and J. Smith. "Wafer defect map similarity search using deep learning in semiconductor manufacturing." In *Proceedings of the International Conference on Semiconductor Manufacturing (ICSM)*, San Francisco, CA, 2024, pp. 123–130. https://doi.org/10.1109/ICSM.2024.1234567.

27. Yan, S., C. Luo*, S. Wang, S. Ding, L. Li, J. Ai, Q. Sheng, Q. Xia, Z. Li, Q. Chen, S. Li, H. Dai, and Y. Zhong. "Virtual metrology modeling for Cvd film thickness with Lasso-Gaussian process regression." In *2023 China Semiconductor Technology International Conference* (CSTIC). | 979-8-3503-1100-6/23/$31.00 ©2023 IEEE | https://doi.org/10.1109/CSTIC58779.2023.10219236.
28. Chen, Y. Z. "Virtual metrology of semiconductor PVD process based on a combination of tree-based ensemble model." *ISA Transactions* 103, no. 1 (2020).
29. Imoto, K., T. Nakai, T. Ike, K. Haruki, and Y. Sato. "A CNN-based transfer learning method for defect classification in semiconductor manufacturing." (2018). https://doi.org/10.1109/TSM.2019.2941752.
30. Krizhevsky, A., I. Sutskever, and G. E. Hinton. "ImageNet classification with deep convolutional neural networks." In *Proceedings of the Advances in Neural Information Processing Systems*, 2012, pp. 1097–1105.
31. https://www2.deloitte.com/content/dam/Deloitte/tw/Documents/technology-media-telecommunications/tw-semiconductor-report-EN.pdf.
32. Wang, J., Y. Ma, L. Zhang, R. X. Gao, and D. Wu. "Deep learning for smart manufacturing: Methods and applications." *Journal of Manufacturing Systems*, 48 (July 2018): 144–156.
33. Indian Council for Research on International Economic Relations (ICRIER), *Globalise to Localise Exporting at Scale and Deepening the Ecosystem are Vital to Higher: Domestic Value Addition in Electronics*, 2022. Available at https://icrier.org/pdf/Globalise_to_Localise.pdf.
34. White paper: ESDM Industry, KPMG in India, 2020.

7 Numerical Simulation-Based Biosensing Performance Exploration of a Cylindrical BioFET Using Machine Learning

Amit Das, Sonam Rewari, Binod Kumar Kanaujia, S. S. Deswal and R. S. Gupta

7.1 INTRODUCTION

Field effect transistors (FETs) are a crucial advancement in electronics, featuring various types for different uses and applications [1–5]. Their adaptability is especially valuable in the biomedical sector, where they serve numerous clinical purposes. FET-based biosensors are utilized for diagnosing and monitoring diseases like Alzheimer's [6], cancer [7], HIV, and cardiovascular diseases (CVD) [8]. However, selectivity and non-unity fill-in factors can introduce minor deviations in real-time measurements.

With scientific and technological progress, FETs have found applications in critical medical devices, such as defibrillators, drug delivery systems, label-free detection of biomolecules, in vivo dosimetry, CVD monitoring, hearing aids, and other implantable devices. Among these, FET-based biosensors (BioFETs) are extensively studied for their potential as economical and dependable alternatives to conventional clinical sensors [9, 10]. They are particularly notable for their high accuracy and linear response [11–13], making them suitable for in vivo dosimetry during radiotherapy [14, 15]. Grenier et al. [16] highlighted the use of FETs in bacterial transformation, which is significant for medical procedures like DNA transfusion and electroporation. Furthermore, Morf et al. [17] described a THz sensor based on an antenna-coupled MOSFET bolometer, useful in biomedical fields for remote detection of hazardous chemicals and material characterization (detection of contrabands without using ionizing radiation) [18, 19].

A biosensor is a tool crafted for the rapid identification of biomolecules. In bioelectronics, BioFETs have risen in popularity due to their exceptional sensitivity and scalability [10]. These biosensors function by utilizing the molecular characteristics, like dielectric constant and charge density, of biochemical substances for

DOI: 10.1201/9781003508304-7

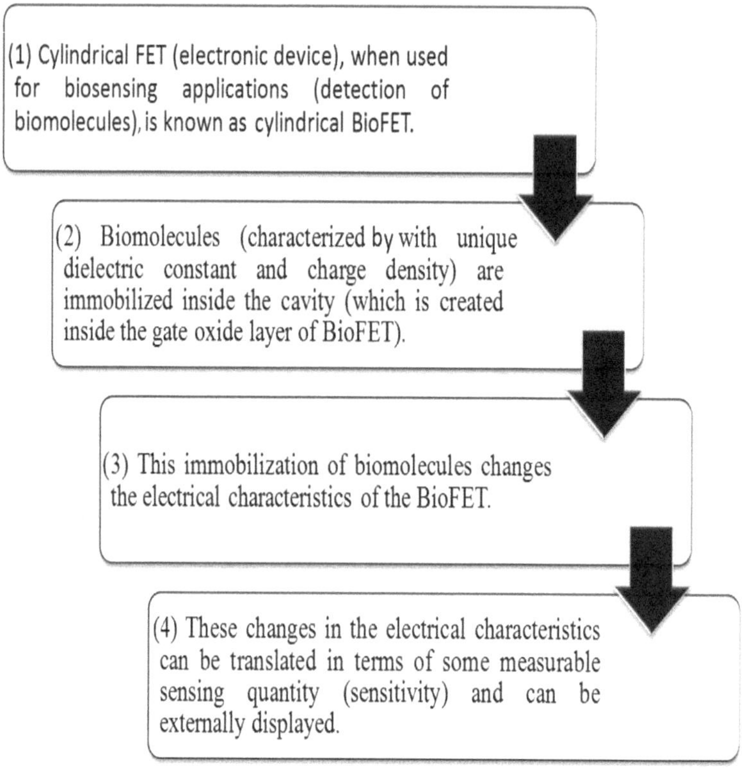

FIGURE 7.1 Fundamental device principle in cylindrical BioFET.

Source: [10].

label-free detection, allowing direct recognition of biomolecules without the need for tagged entities [10, 20]. The primary detection mechanism in a BioFET centers on the dielectric modulation of sensing parameters [20, 21]. Figure 7.1 outlines the fundamental principle as well as basic detection mechanism of a BioFET.

Threshold voltage serves as a primary sensing parameter in BioFETs, complemented by metrics like subthreshold slope, OFF current, I_{ON}/I_{OFF} ratio, and transconductance for sensitivity evaluation. Within BioFETs, a cavity is engineered in the gate oxide layer to effectively immobilize biomolecules. This structural feature alters electric potential and field distribution upon biomolecule immobilization, thereby inducing changes in the threshold voltage. These changes reflect the biosensor's sensitivity to specific biomolecules, facilitating detection through the modulation of dielectric properties. Effective biosensors exhibit heightened sensitivity, manifesting as significant threshold voltage shifts in response to a biomolecule's presence. Diverse BioFET variants have been reported and described in literature [22–26]. Ahangani et al. [27] and Li et al. [28] have explored sensitivity through threshold voltage dynamics and other parameter variations. Recently, Pratap et al. [29] and Chakraborty et al. [20] pioneered a novel cylindrical BioFET design using junctionless transistors, employing diverse sensing metrics for biomolecule detection. Gao et al. [30] demonstrated nanowire FETs operating in the subthreshold range for protein- and pH-sensing applications. Gautam et al. [5] and Pour et al. [31] highlighted cylindrical FETs for gas-sensing applications. Biochemical species size spans from millimeters to femtometers [10, 29]. Dam et al. [32] achieved robust sensitivity with a FET-based glucose biosensor targeting glucose oxidase detection. Recently, Koshti et al. [12] introduced a FET-based biosensor for detecting the COVID-19 virus.

Over the past decade, researchers have explored a range of BioFET designs presented by various authors. These biosensors have been investigated with different structural configurations, oxide materials, and work functions. However, there has been no prior report on enhancing sensitivity through doping optimization to modulate the threshold voltage in these sensors. The primary focus of this chapter is to optimize doping profiles in the source, drain, and channel regions to achieve a significantly enhanced sensitivity in BioFETs. In practical applications, it is typically expected for the cavity within BioFETs to remain partially filled due to unavoidable circumstances. Therefore, it is essential to examine how partially filled cavities impact the sensitivity of the biosensor. The fill-in factor, which denotes the fraction of the cavity volume occupied by biomolecules, plays a crucial role in this context [10]. Despite its importance, the comprehensive study of fill-in factor effects on BioFETs' sensitivity has been lacking in the related literature [33]. Hence, the second objective of the research in this chapter is to comprehensively investigate the influence of fill-in factor on threshold voltage sensitivity in the context of biosensing applications [33, 34].

Traditional biosensors often face challenges regarding sensitivity, specificity, and adaptability. However, machine learning (ML) algorithms can effectively address these issues and significantly improve a biosensor's performance. ML, a discipline within artificial intelligence (AI), empowers computers to learn from data and make predictions or decisions based on that information. In biosensing, ML algorithms can analyze complex data generated by biosensors, enhance accuracy, and automate decision-making processes [35]. By harnessing ML, biosensors can be transformed into more potent instruments for detecting and monitoring diseases, pollutants, and toxins [36]. This discussion briefly centers on two key aspects: utilizing machine learning for data analysis and applying ML in sensor design. Finally, it briefly addresses the challenges and opportunities for advancing ML-based biosensors. Moreover, ML enables biosensors to adapt to varying environmental conditions and optimize performance through continuous learning. This capability is crucial for real-world applications, where environmental factors may fluctuate with time. ML-driven biosensors also promise to integrate with internet of things (IoT) platforms, facilitating remote monitoring and real-time data transmission for enhanced usability and accessibility [35]. Figure 7.2 shows the basics of the principle behind the integration of ML and BioFET engineering.

A cylindrical FET has been selected here for its ability to effectively control charge carrier flow and maintain electrostatic integrity, owing to its surrounding gate design. The first section briefly reviews the literature, identifies research gaps, and outlines the scope of the current study. The second section elaborates on

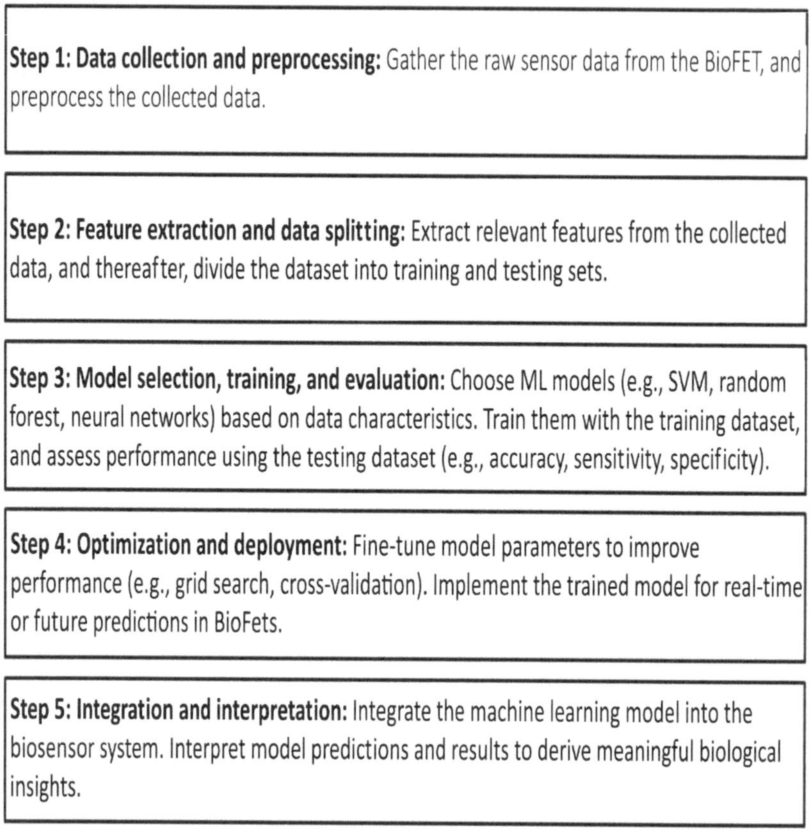

FIGURE 7.2 Basic flowchart explaining the implementation of machine learning in studying BioFET.

BioFET calibration, fabrication processes, simulator setup, and the methodology employed. In the third section, significant findings and corresponding graphs are discussed, focusing on the impact of doping, temperature variations, and fill-in factors on threshold voltage sensitivity. Finally, the fourth section summarizes key findings, discusses potential applications, and underscores the innovative aspects of the current research.

7.2 DEVICE AND SIMULATOR SPECIFICATIONS

7.2.1 BioFET Calibration and Characteristics

The cylindrical BioFET was calibrated with standard values to optimize sensitivity. A polysilicon gate with a work function of 4.96 eV was chosen for its fabrication simplicity and wide operating temperature range compared to metallic gates. The bias supply was limited to 1 V, which is the standard for nanoscale BioFETs.

The electron affinity was set at 4.17 eV. The calibrated values of energy bandgap, effective density of state in valence, and conduction band at room temperature are 1.08 eV, $1.04 \times 10^{19}/cm^3$, and $2.8 \times 10^{19}/cm^3$, respectively [37, 38]. Figures 7.3a and 7.3b present a 3D and 2D view of the proposed BioFET. Opting for a symmetric two-sided cavity design over a single-sided configuration offers advantages such as improved fill-in probability, reduced power consumption, and enhanced current sensitivity [39, 40]. It is feasible to embed the cavity within the SiO_2 gate oxide layer [29]. Detailed fabrication steps for the BioFET are depicted in Figure 7.3c using a flowchart format [28, 41–43].

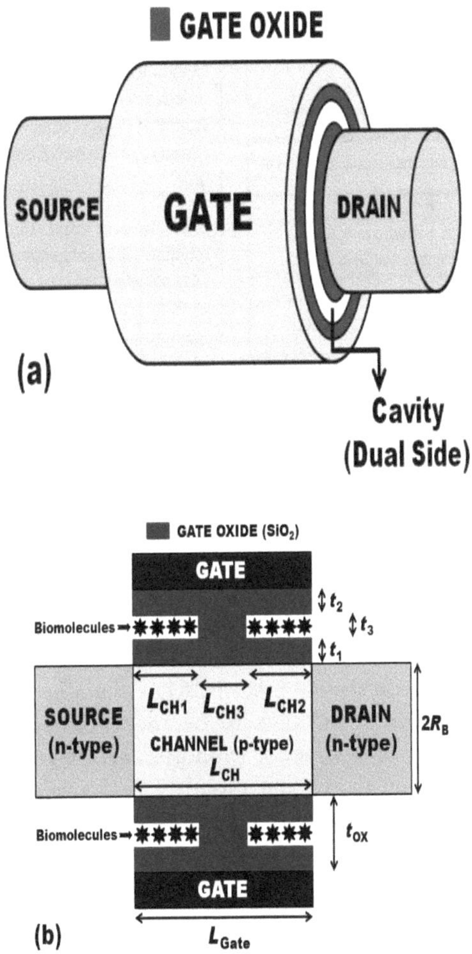

FIGURE 7.3 (a) 3D view, (b) 2D view, and (c) fabrication process flowchart, illustrating the basic fabrication steps of a cylindrical BioFET.

Source: [44].

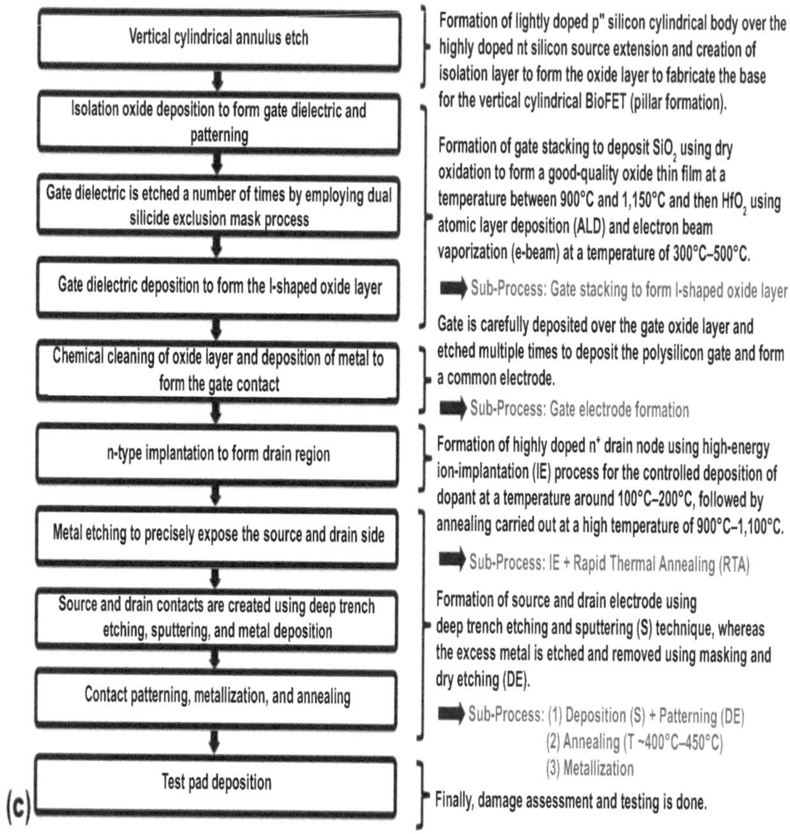

FIGURE 7.3 (Continued)

7.2.2 Pre-Setup Specifications of TCAD

The numerical investigation of the cylindrical BioFET is analyzed using the values listed in Table 7.1 on the ATLAS TCAD simulator [45]. To imitate real-time conditions, related models have to be incorporated during simulation: the Shockley-Read-Hall (SRH) model supports carrier generation and recombination, the concentration-dependent mobility (CONMOB) model relates the impurity profile to low field mobility at room temperature, and the parallel electric field–dependent mobility (FLDMOB) model accounts for field-dependent mobility [46]. To solve the nonlinear and complex differential equations, the Newton-Gummel method is employed, which integrates both decoupled and coupled iterations [46, 47].

7.2.3 Computational Methodology

Biomolecules are defined by their specific charge densities (ρ) and dielectric constants (K_{bio}) [20, 26, 47]. For example, non-hybridized DNA carries a negative charge

TABLE 7.1
Structural Parameters

Parameters		**Value**		
Channel length (L_{CH})		30 nm		
Source/channel/drain radius (R_B)		10 nm		
	Source	n-type		
Material	Channel	p-type	Silicon (Si)	
	Drain	n-type		
Gate work function [48, 49]		4.96 eV		
Channel doping (N_C)		$10^{10}/cm^3$		
Source doping (N_S)		1×10^{13} to $5 \times 10^{20}/cm^3$		
Drain doping (N_D)		1×10^{13} to $5 \times 10^{20}/cm^3$		
Source/drain length		10 nm		
Oxide layer thickness (t_{ox})		6 nm		
Cavity thickness		$t_1 = 1$ nm	$t_2 = 4$ nm	$t_3 = 1$ nm
Cavity length		$L_{CH1} = L_{CH2} = 14$ nm	$L_{CH3} = 2$ nm	
Gate oxide		Silicon dioxide (SiO_2)		
Biasing		$V_{DS} = (0–1)$ V	$V_{GS} = (0–1)$ V	
Oxide length		30 nm		
Biomolecules				
Neutral biomolecules				
Uricase ($K_{bio} = 1.54$) [29]		Streptavidin ($K_{bio} = 2.1$) [50]		
Protein ($K_{bio} = 2.5$) [29]		Biotin ($K_{bio} = 2.63$) [51]		
ChOX ($K_{bio} = 3.3$) [29]		APTES ($K_{bio} = 3.57$) [50]		
Hydroprotein ($K_{bio} = 5$) [52]		Keratin ($K_{bio} = 8$) [53]		
Charged biomolecules				
DNA [54]		Amino acids [55]		
$K_{bio} = 5$; $\rho = -1 \times 10^{11}/cm^2$		$K_{bio} = 5$; $\rho = +1 \times 10^{11}/cm^2$		

Note: ChOX, Cholesterol Oxidase; APTES, 3-Aminopropyltriethoxysilane.

and shows variable charge density and dielectric constant [54]. In contrast, neutral biomolecules such as uricase, zein, and streptavidin [26] have only dielectric constants. The various test biomolecules utilized in this study are listed in Table 7.1. This study includes a range of biomolecules with different charge densities to examine sensitivity pattern variations. By considering both charged and neutral biomolecules, the investigation aims to deliver a comprehensive and realistic analysis of biosensor sensitivity.

When the cavity is empty, it contains air with a dielectric constant of $K_{air} = 1$. Introducing a test biomolecule (with $K_{bio} \neq 1$) into the cavity leads to changes in the threshold voltage and other device characteristics. This variation occurs because the biomolecule alters the gate oxide capacitance and the lateral electric field, thus modifying the potential distribution along the channel [10]. The relative shift in the

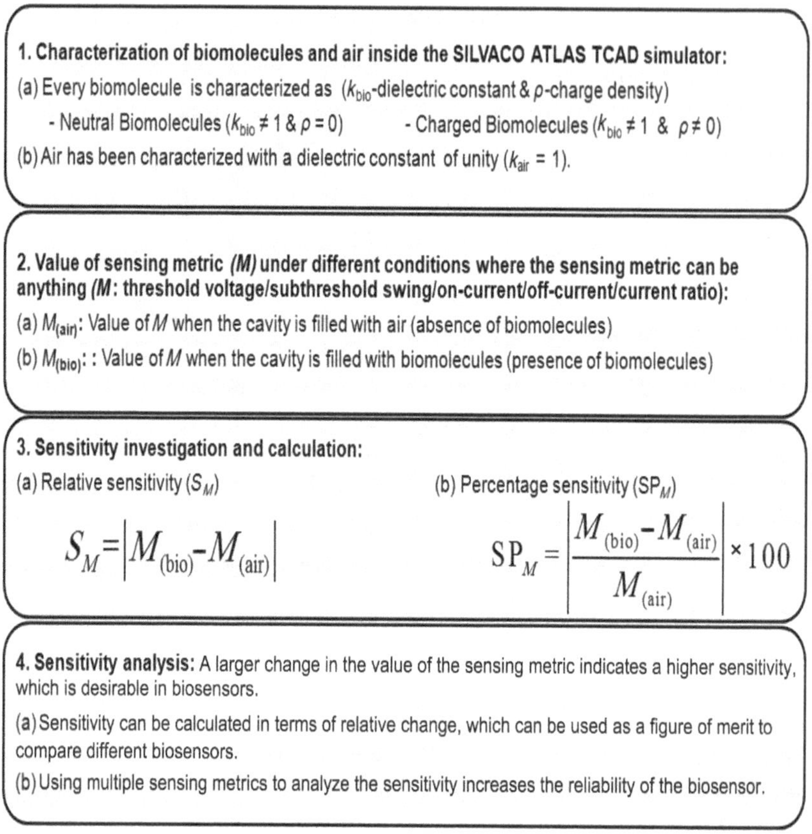

FIGURE 7.4 Flowchart depicting the summary of methodology used for sensitivity analysis.

threshold voltage serves as a qualitative gauge in the biosensing investigation of a BioFET. It's important to note that while biomolecules impact the oxide capacitance and thereby affect the lateral electric field, doping influences the horizontal electric field by adjusting the potential barrier at the source-channel and drain-channel junctions [56, 57]. The numerical methodology employed for sensitivity analysis [10, 20] is concisely represented in the flowchart shown in Figure 7.4.

7.3 RESULTS AND DISCUSSION

7.3.1 Biomolecules Detection for Clinical Applications

Early detection of several chronic diseases can be effectively achieved using cylindrical BioFET. These include conditions such as ovarian or breast cancer, Alzheimer's disease, and hepatocellular carcinoma (HCC), where specific biomolecules play a critical role in disease screening. BioFET can detect uricase ($K_{bio} = 1.54$) [29],

TABLE 7.2
Shift in Threshold Voltage for Different Biomolecules

Biomolecules		Shift in Threshold Voltage (ΔV_t)	
Uricase		39.0281 mV	
Streptavidin		74.5910 mV	
Cholesterol Oxidase		134.914 mV	
Aminopropyl Triethoxysilicane		141.048 mV	
$N_S = N_D = 10^{19}/cm^3$	$L_1 = L_2 = 14$ nm	$N_C = 10^{10}/cm^3$	$V_{DS} = 0.5$ V

essential for preventing uric acid nephropathy, and streptavidin ($K_{bio} = 2.1$) [50], used in immunoassays and blotting. Other detectable biomolecules include cholesterol oxidase ($K_{bio} = 3.3$) [29], commonly used in biocatalysis, and aminopropyl triethoxysilane ($K_{bio} = 3.57$) [50], crucial for the silanization process. Table 7.2 details the observed shifts in threshold voltage when these biomolecules are localized within the nanogap cavity of the biosensor (BioFET).

7.3.2 Effect of Doping

Increasing the doping of source and drain in an identical pattern will lead to two primary effects: Effect 2 and Effect 1. When the doping of source and drain are increased at an equal rate from a low to high value, a major part of the depletion region will penetrate inside the channel [37] rather than the source (source-channel junction) or drain (drain-channel junction), which is basically Effect 1. The aggregate depletion layer width keeps on reducing with increased doping [58], which is basically Effect 2. So both Effect 1 and Effect 2 will be present when the doping of source and drain are increased from a low to high value. Effect 2 is more dominant at low doping, which leads to reduction of the depletion layer width, thereby increasing the effective channel length. Effect 1 is more dominant at high doping, which leads to penetration of the depletion layer inside the channel, resulting in a decrease of the effective channel length. These two effects are shown in the diagrammatic flowchart in Figure 7.5. Doping notation s_t_u indicate doping of the source ($1 \times 10^s/cm^3$), channel ($1 \times 10^t/cm^3$), and drain ($1 \times 10^u/cm^3$) in the same order, and the S_C_D notations indicate source_channel_drain, respectively. Values of doping D_A–D_H are specified in Table 7.3.

Figure 7.6 depicts the variation in threshold voltage sensitivity across different biomolecules. Initially, the threshold voltage increases due to the longer effective channel length at lower doping levels (Effect 2), necessitating a higher gate voltage to activate the FET. This increase in gate voltage results in a higher threshold voltage, leading to enhanced sensitivity at lower gate voltages. Conversely, at higher doping levels, the effective channel length decreases (Effect 1), requiring a lower gate voltage (threshold voltage) for device activation. Therefore, sensitivity decreases as doping levels increase. The threshold voltage shows an initial increase followed

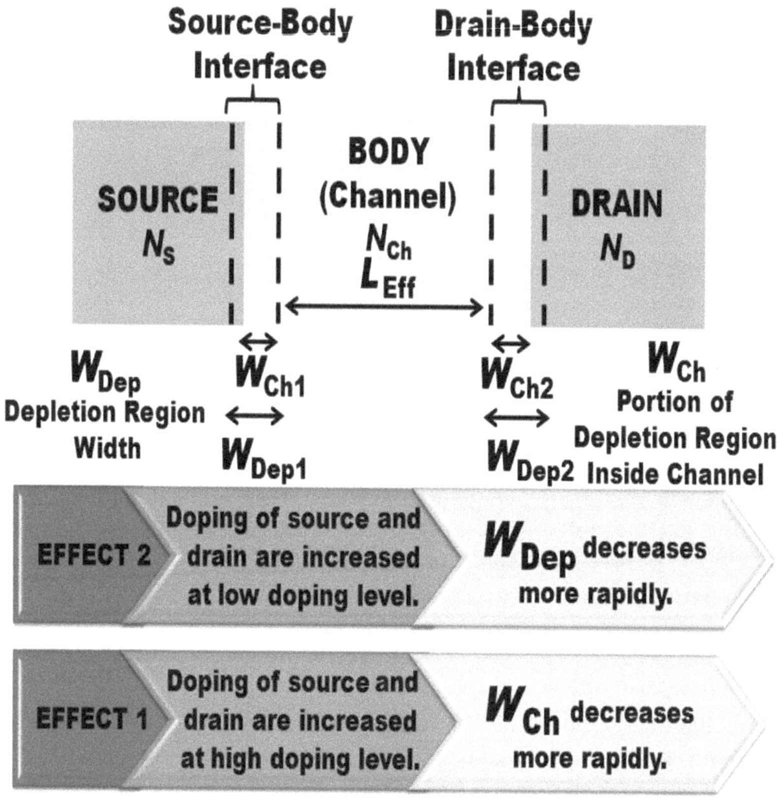

FIGURE 7.5 Visualization of the effect of doping on effective channel length.

by a decrease as the doping levels of the source and drain increase uniformly, as illustrated in the inset of Figure 7.6. Notably, when biomolecules are present within the cavity at low doping levels, the threshold voltage changes by more than ~400%, indicating substantial sensitivity, crucial for biosensing applications.

While achieving high percentage sensitivity at low doping levels is beneficial, the device's conductivity reduces under such conditions, imposing limitations on using low doping in BioFETs. Thus, a balance between sensitivity and conductivity must be struck through optimized doping levels to achieve optimal performance. Additionally, the presence of charged biomolecules affects the value of the threshold voltage. Positively charged biomolecules attract electrons from the substrate, creating a channel at a lower gate voltage and reducing the threshold voltage compared to neutral biomolecules. Conversely, negatively charged biomolecules increase the threshold voltage. Therefore, the sensitivity of the threshold voltage exhibits a similar trend for charged biomolecules as observed with neutral biomolecules. Understanding these dynamics is crucial for tailoring BioFET designs to specific biomolecular interactions and optimizing their performance in diverse biological and environmental applications. It is worth noting that the relative change in V_t is significant at higher

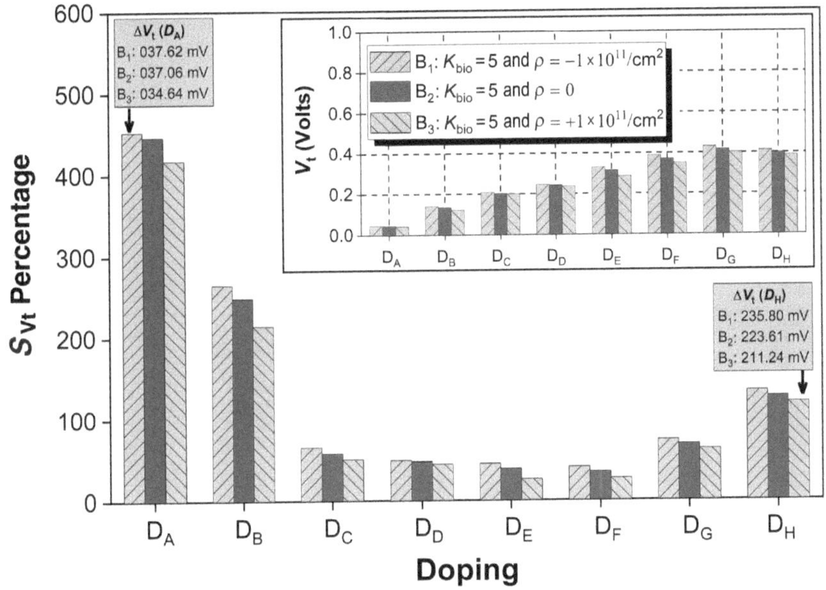

FIGURE 7.6 Threshold voltage sensitivity (S_{Vt}) at different doping for different test biomolecules. ([**INSET**] Shows the variation of threshold voltage at different doping for different test biomolecules.)

doping levels and minimal at lower doping levels, which is the opposite of the trend observed in the percentage change in sensitivity.

Table 7.3 provides data on the subthreshold slope (SS) and subthreshold slope sensitivity (S_{SS}) across different symmetric doping levels. The subthreshold slope is derived from the log (I_{DS})-V_{GS} curve (transfer characteristics) and inversely correlates with the depletion capacitance. At lower doping levels, where Effect 2 predominates, the depletion width decreases as doping increases, resulting in higher depletion capacitance and lower subthreshold slope (SS shows a nonlinear inverse relationship with depletion capacitance). Conversely, at higher doping levels, dominated by Effect 1, the subthreshold slope increases with doping. A higher subthreshold slope indicates more significant variability in drain current for the same gate voltage range, showing a more pronounced change at higher doping levels. Local minima in the subthreshold slope pattern are evident in Table 7.3. Positively charged biomolecules marginally increase the ON current, while negatively charged biomolecules slightly decrease it compared to neutral biomolecules. This increase in ON current enhances subthreshold characteristics, leading to an increase in subthreshold slope, although the relative change reduces. Importantly, the threshold voltage exhibits a greater relative change than the subthreshold slope, highlighting its superior sensitivity as a metric for analyzing biosensor performance. Understanding these dynamics is essential for advancing FET-based biosensors toward enhanced sensitivity and reliability in complex biological and environmental contexts. Optimization strategies can thus leverage these insights to refine sensor design and performance.

TABLE 7.3
SS and S_{SS} Variation at Different Symmetric Doping

Doping Value			$K_{bio} = 5$		
			ρ	ρ	ρ
$N_S\_N_C\_N_D$ (cm^{-3})			-1×10^{11} cm^{-2}	0	$+1 \times 10^{11} cm^{-2}$
D_A	$10^{13}\_10^{10}\_10^{13}$	SS	71.943	72.016	72.949
		S_{SS}	25.11756	25.0421	24.07005
D_B	$10^{14}\_10^{10}\_10^{14}$	SS	71.933	71.957	71.999
		S_{SS}	24.08303	24.05791	24.0545
D_C	$10^{15}\_10^{10}\_10^{15}$	SS	71.413	71.417	71.476
		S_{SS}	24.04949	24.04556	23.98249
D_D	$10^{16}\_10^{10}\_10^{16}$	SS	71.2799	71.28	71.281
		S_{SS}	23.82677	23.82673	23.82566
D_E	$10^{17}\_10^{10}\_10^{17}$	SS	71.206	71.209	71.212
		S_{SS}	23.67215	23.66915	23.66604
D_F	$10^{18}\_10^{10}\_10^{18}$	SS	71.144	71.146	71.149
		S_{SS}	23.96603	23.964	23.96058
D_G	$10^{19}\_10^{10}\_10^{19}$	SS	77.467	77.612	77.761
		S_{SS}	37.11498	36.99703	36.87624
D_H	$10^{20}\_10^{10}\_10^{20}$	SS	93.946	94.072	94.226
		S_{SS}	44.17691	44.10204	44.01065

SS: $(mV/decade)^{-1}$ S_{SS}: (Percentage)

$$S_{SS} = \left| \frac{SS_{(air)} - SS_{(bio)}}{SS_{(air)}} \right| \times 100 \tag{7.1}$$

Figure 7.7 illustrates the changes in sensitivity of the threshold voltage, accompanied by an inset showing the sensitivity of the subthreshold slope, as the doping levels of (i) the drain (with constant source and channel doping) and (ii) the source (with constant drain and channel doping) are increased from a reference doping level (D_{RD}: $N_C = 10^{10}/cm^3$ and $N_S = N_D = 10^{19}/cm^3$). The chosen reference doping level is high, so further increasing either source or drain doping enhances both threshold voltage sensitivity and subthreshold slope sensitivity due to the prevailing Effect 2 at high doping. Notably, sensitivity is significantly higher when drain doping exceeds source doping. Detailed doping levels (D_{D1}–D_{D3} and D_{S1}–D_{S3}) are outlined in Table 7.4, which also illustrates the observed patterns of subthreshold slope and threshold voltage at different asymmetric doping, revealing the local minima at the reference doping level. Figure 7.8 demonstrates the sensitivity variations observed with different neutral biomolecules. An increase in the dielectric constant of biochemical species leads to a corresponding increase in the oxide capacitance within the cavity. This increased gate oxide capacitance strengthens

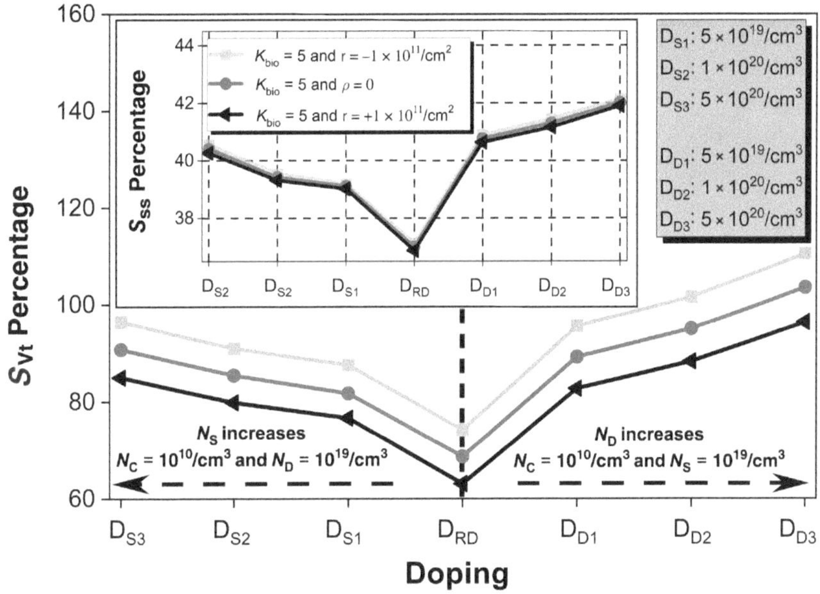

FIGURE 7.7 Threshold voltage sensitivity (S_{Vt}) at different asymmetric doping for different test biomolecules. ([**INSET**] Shows the variation of subthreshold slope sensitivity at different asymmetric doping for different test biomolecules.)

the interaction between charge carriers and the gate [20]. As a result, both the threshold voltage and subthreshold slope exhibit more pronounced changes at higher values of K_{bio} (relative to the empty cavity scenario), indicating increased sensitivity of the biosensor to biomolecules with higher K_{bio}. The inset table in Figure 7.8 outlines the variations in threshold voltage sensitivity and subthreshold slope sensitivity for various charged biomolecules, highlighting a stronger sensitivity toward negatively charged biomolecules due to enhanced gate-to-channel coupling. Increasing the dielectric constant of biomolecules enhances the gate oxide capacitance, intensifying the interaction between charge carriers and the gate. This amplifies the sensitivity of FET-based biosensors, particularly toward biomolecules with higher dielectric constants, thereby enhancing detection capabilities in diverse biochemical environments. Understanding these dynamics is pivotal for tailoring BioFETs to effectively detect biomolecules with varying charges and dielectric characteristics.

Figure 7.9 illustrates the changes in the I_{ON}/I_{OFF} ratio for hydroprotein across different doping levels. The inset provides specific comparisons of the I_{ON}/I_{OFF} ratio at low and high doping levels for various biomolecules. Increasing the source doping enhances both the ON-state current (I_{ON}) and OFF-state current (I_{OFF}) [59]. This increase in I_{ON} primarily contributes to a higher overall I_{ON}/I_{OFF} ratio. However, at higher doping levels, the ON-state current begins to decrease due to significant degradation in electron mobility.

TABLE 7.4
V_t and SS Variation at Different Asymmetric Doping

Doping Value	$N_S\_N_C\_N_D$ (cm^{-3})		ρ $-1*10^{11}$ cm^{-2} ($K_{bio} = 5$)	ρ 0 ($K_{bio} = 5$)	ρ $+1*10^{11}$ cm^{-2} ($K_{bio} = 5$)
D_{S3}	$5*10^{20}\_1*10^{10}\_1*10^{19}$	SS	84.7319	84.933	85.1551
		V_t	424.49	412.21	399.754
D_{S2}	$1*10^{20}\_1*10^{10}\_1*10^{19}$	SS	83.8564	84.015	84.1602
		V_t	427.493	415.15	402.624
D_{S1}	$5*10^{19}\_1*10^{10}\_1*10^{19}$	SS	82.593	82.705	82.8195
		V_t	429.305	416.51	402.73
D_{RD}	$1*10^{19}\_1*10^{10}\_1*10^{19}$	SS	77.4668	77.612	77.7609
		V_t	430.309	416.80	402.911
D_{D1}	$1*10^{19}\_1*10^{10}\_5*10^{19}$	SS	83.1755	83.341	83.5154
		V_t	419.28	405.74	391.808
D_{D2}	$1*10^{19}\_1*10^{10}\_1*10^{20}$	SS	84.9219	85.126	85.3389
		V_t	416.489	402.94	388.99
D_{D3}	$1*10^{19}\_1*10^{10}\_5*10^{20}$	SS	86.6484	86.810	86.9765
		V_t	412.482	399.28	385.3

N_S / RD / N_D

(**Case i**) N_D varies (N_C & N_S are constant) (**Case ii**) N_S varies (N_C & N_D are constant)

SS : (mV/decade)$^{-1}$ V_t : mV

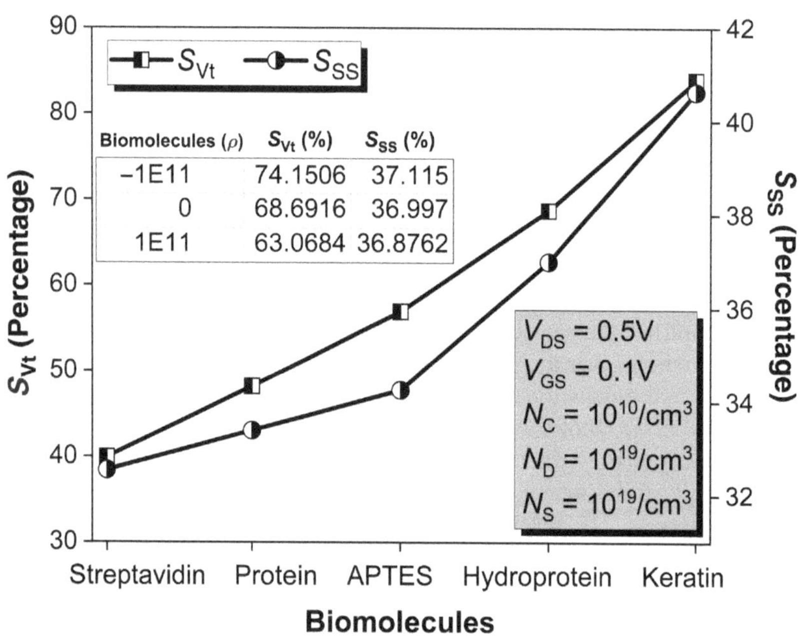

FIGURE 7.8 Threshold voltage sensitivity and subthreshold slope sensitivity in the presence of different neutral biomolecules.

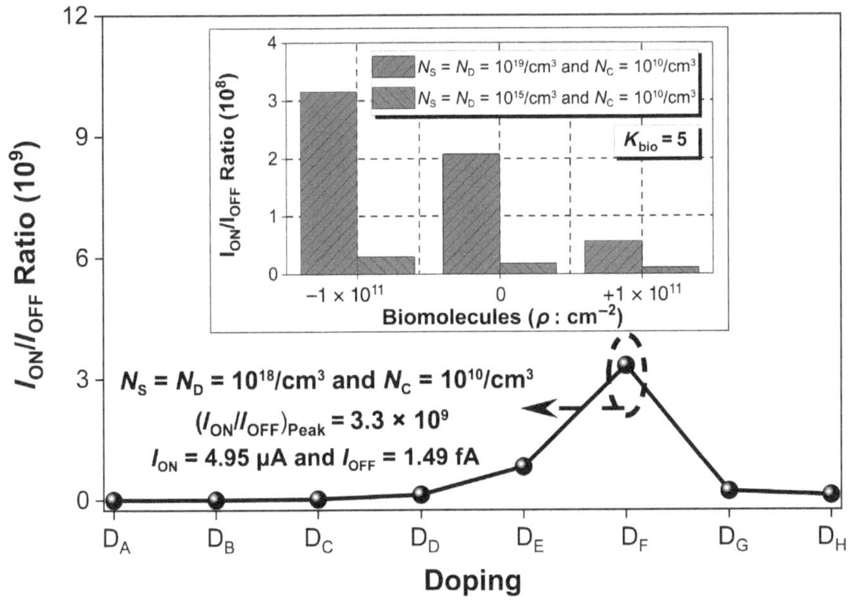

FIGURE 7.9 I_{ON}/I_{OFF} ratio at different doping for hydroprotein. ([**INSET**] Shows the I_{ON}/I_{OFF} ratio variation for different charged biomolecules.)

Figure 7.10 depicts the sensitivity analysis of the threshold voltage and subthreshold slope across varying channel doping levels. Increasing the channel doping causes a slight rise in the threshold voltage because more gate voltage is needed to deplete the channel fully. This increase in doping increases the number of holes that must be depleted before electrons can flow from source to drain under applied voltage. However, the sensitivity of the threshold voltage, reflecting its relative change, decreases with higher N_C. Similarly, the subthreshold slope and its sensitivity decline with increased channel doping. High N_C levels reduce both the ON-state and OFF-state currents, leading to a gentler slope in the I_{DS}–V_{GS} plot within the subthreshold region. Consequently, as N_C rises, both the subthreshold slope and its sensitivity reduces. These findings emphasize the critical role of channel doping and dielectric properties in tuning biosensor performance. By optimizing these parameters, FET-based biosensors can achieve superior detection capabilities, essential for various biomedical and environmental applications.

Figures 7.11 and 7.12 depict how the drain current changes with gate and drain voltages, respectively, across the immobilization of different neutral biomolecules. The inset in these figures show similar changes for various charged biomolecules. The relative difference in drain current with biomolecule presence (compared to an empty cavity) increases as K_{bio} rises due to enhanced coupling between the channel and gate. Higher K_{bio} values lead to more significant reductions in drain current, indicating more considerable changes at higher dielectric constants [20]. Negatively charged biomolecules further enhance this coupling, resulting in increased sensitivity

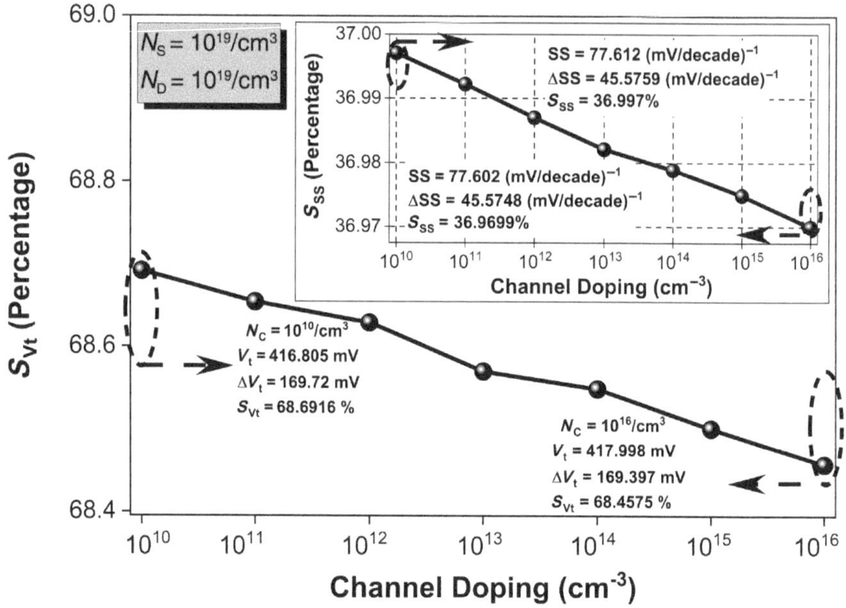

FIGURE 7.10 Threshold voltage sensitivity at different channel doping for hydroprotein. ([**INSET**] Shows the subthreshold slope sensitivity at different channel doping for hydroprotein.)

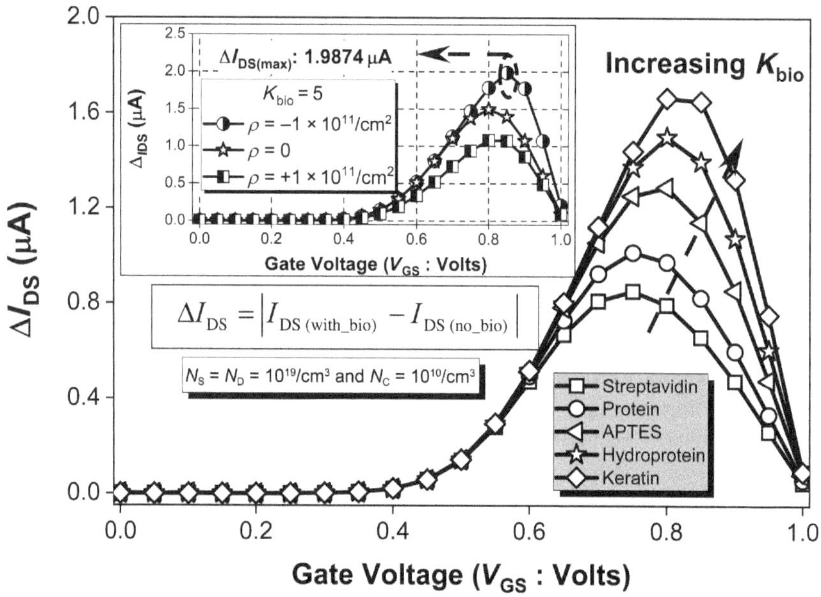

FIGURE 7.11 Change in drain current as a function of gate voltage at V_{DS} = 0.5 V for different neutral biomolecules. ([**INSET**] Shows the change in drain current as a function of gate voltage at V_{DS} = 0.5 V for different charged biomolecules.)

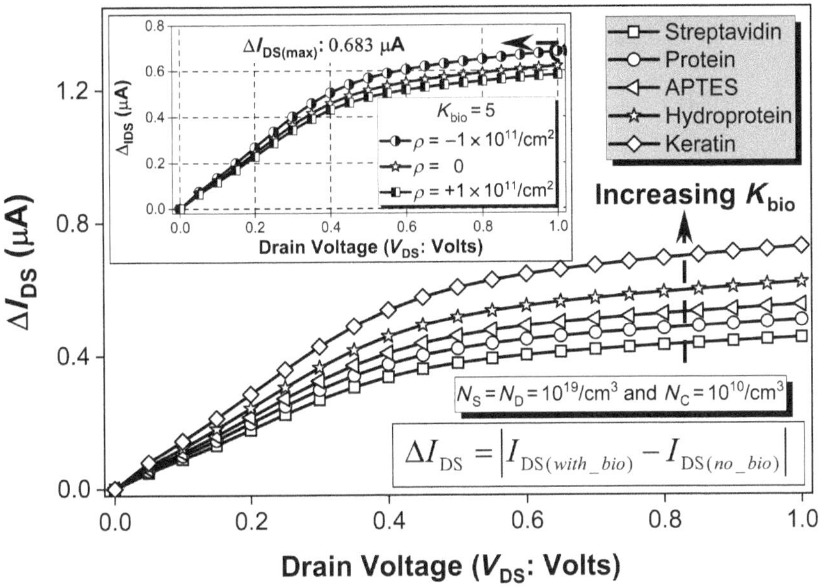

FIGURE 7.12 Change in drain current as a function of drain voltage at V_{GS} = 0.5 V for different neutral biomolecules. ([**INSET**] Shows the change in drain current as a function of drain voltage at V_{GS} = 0.5V for different charged biomolecules.)

compared to neutral biomolecules. Notably, the cylindrical BioFET demonstrates higher sensitivity to gate voltage than drain voltage in the presence of biomolecules. This knowledge also paves the way for creating versatile BioFET platforms capable of reliably detecting and measuring diverse biomolecules across different biological matrix species, maximizing device performance for specific biomarker detection.

Figure 7.13 depicts how the threshold voltage sensitivity and subthreshold swing sensitivity vary with drain bias at high doping levels, while the inset shows these sensitivities at low doping levels. Both sensitivities increase as the high drain voltage generates a greater number of charge carriers. This increase in charge carriers leads to more significant changes in threshold voltage and subthreshold slope when biomolecules are present. As a result, the device's sensitivity to biomolecules is high at higher drain voltages, owing to a stronger electric field across the channel. However, it is important to note that the drain voltage cannot be increased indefinitely, as excessively high drain voltage in a short-channel device may cause permanent damage. This knowledge also paves the way for creating versatile BioFET platforms capable of reliably measuring diverse biomolecules across different biological matrices, maximizing device performance for specific biomarker detection. Figure 7.14 illustrates the potential distribution across the cylindrical BioFET channel for various neutral biomolecules, with the inset displaying the minimum potential across the channel at different symmetric doping levels. The increase in channel potential curvature

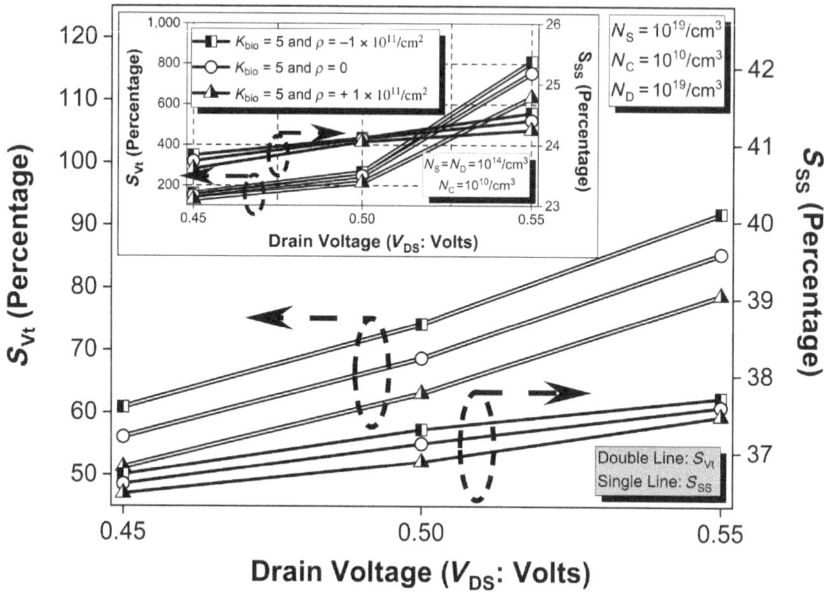

FIGURE 7.13 Threshold voltage sensitivity and subthreshold slope sensitivity at various drain voltages for different biomolecules at high doping. ([**INSET**] Shows the threshold voltage sensitivity and subthreshold slope sensitivity at various drain voltages for different biomolecules at low doping.)

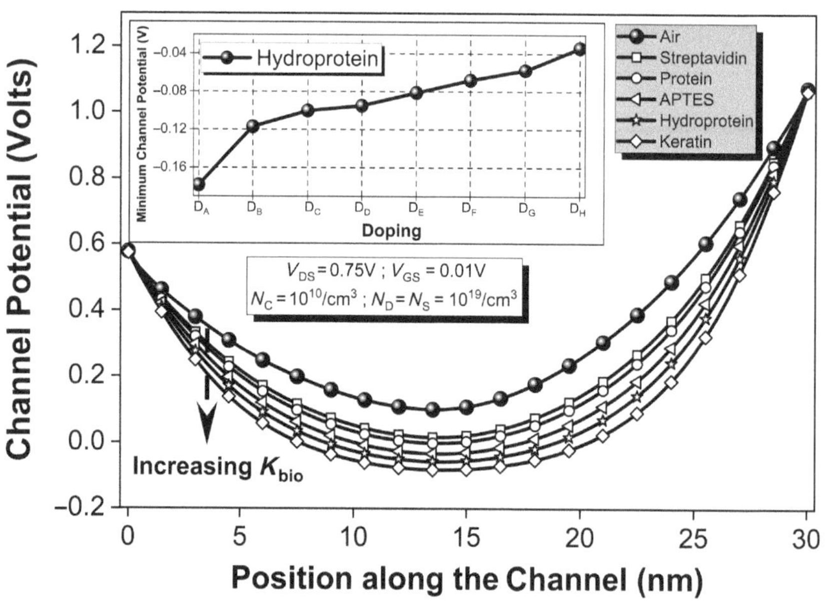

FIGURE 7.14 Potential across the channel in the presence of different biomolecules. ([**INSET**] Shows minimum channel potential for hydroprotein.)

with higher K_{bio} suggests that the biosensor is more sensitive to biomolecules with higher dielectric constants, supporting the results of this study. The greater relative change in channel potential is a critical parameter for evaluating the biosensing performance of the cylindrical BioFET. Notably, increasing the gate voltage reduces the potential curvature, so the channel potential is plotted at a low gate voltage of 10 mV to emphasize the effect of biomolecules on the curvature. These results highlight the necessity of carefully balancing drain and gate voltages to optimize sensor performance while avoiding device damage. Understanding the changes in potential distribution across the channel further underscores the importance of dielectric properties in enhancing biosensor sensitivity. Such insights are crucial for designing robust and highly sensitive BioFETs tailored for specific biomedical and environmental applications.

Figure 7.15 shows how temperature affects the threshold voltage sensitivity, with an inset for different charged biomolecules. Increased temperature boosts charge carrier generation [60], lowering the threshold voltage and increasing its relative change, enhancing sensitivity to biomolecules. It is crucial for FETs to choose the right gate oxide, like SiO_2 for its stability and compatibility. Integrating high-K dielectrics with SiO_2 reduces fringing fields [18] but sacrifices sensitivity and adds complexity. Figure 7.16 compares sensitivity in the cylindrical BioFET with and without stacked gate oxide for various biomolecules, highlighting sensitivity enhancement versus fabrication complexity.

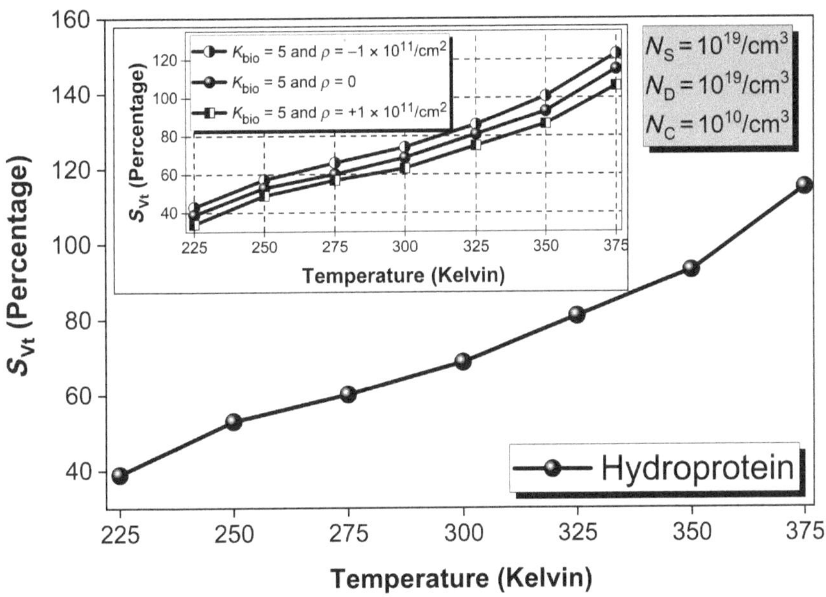

FIGURE 7.15 Threshold voltage sensitivity variation at different temperatures for hydroprotein. ([**INSET**] Shows the threshold voltage sensitivity variation at different temperatures for different charged biomolecules.)

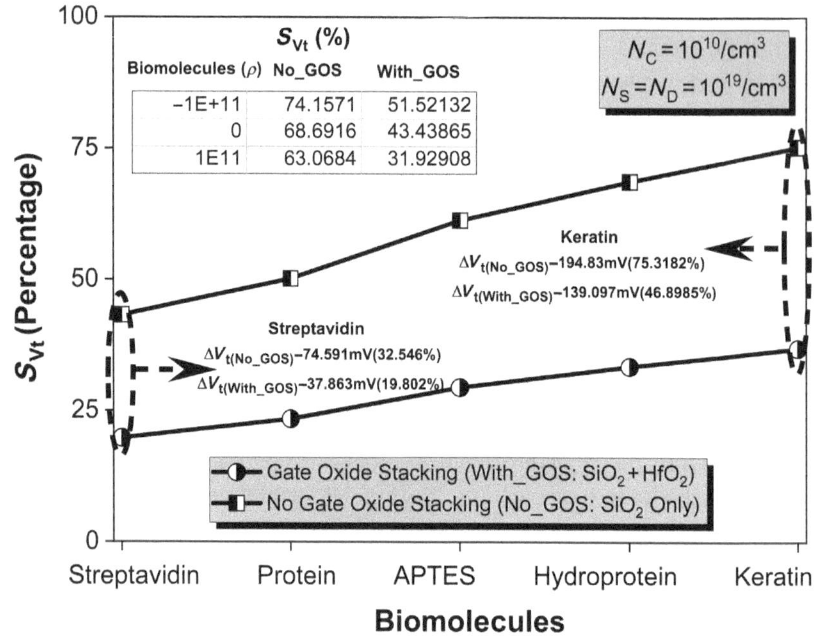

FIGURE 7.16 Threshold voltage sensitivity with and without gate oxide stack for different neutral biomolecules.

Table 7.5 presents the impact of biomolecule location at various fill-in factors. The cavity around the source and drain is divided into four equal parts, allowing different test biomolecules to be examined under various conditions. Case 5 has a larger fill-in factor than Case 4, yet Case 4 shows higher sensitivity. This trend is similarly observed between Case 3 and Case 5, as well as between Case 10 and Case 11, among others.

Typically, a higher fill-in factor increases sensitivity by immobilizing a greater number of biomolecules within the cavity [10]. However, the specific location of biomolecules within partially filled cavities notably impacts sensitivity. Case 4, for example, despite having the same fill-in factor as Case 1, exhibits higher sensitivity because biomolecules are situated closer to the source. This proximity enhances their influence on the potential hill at the source-channel junction, thereby affecting threshold voltage more significantly. Consequently, sensitivity is maximized when biomolecules are positioned nearer to the source [61]. Similar sensitivity trends are observed in comparisons between Case 5 and Case 10, Case 11 and Case 14, and other analogous scenarios.

Figures 7.17 and 7.18 depict a comparative analysis of the cylindrical BioFET biosensor with various existing variants of biosensors. It is evident from the figures that the threshold voltage shift (sensitivity) increases proportionally with the widening of the cavity, allowing for a greater number of biomolecules to be immobilized within it. Consequently, the biosensor shows a more pronounced change in threshold voltage with larger cavity dimensions. Figure 7.18 specifically compares the sensitivity of our

TABLE 7.5
Effect of Fill-In Factor and Location of Different Biomolecules on S_{Vt}

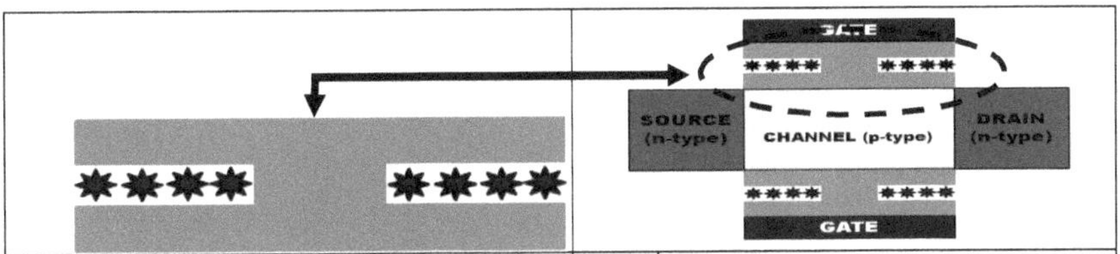

S. No.	Cavity toward the Source End (L_{CH1})		Oxide Layer (L_{CH3})	Cavity toward the Drain End (L_{CH2})		Fill-In Factor	Threshold Voltage Sensitivity (S_{Vt}: mV), Increasing K_{bio} → Biomolecules A	B	C	D	E	F
	Cavity on Both Sides Are Divided into Two-Two Equal Parts											
1.	□	□	▒	□	■	0.25	0.006	0.095	0.597	1.409	1.74	2.027
2.	□	□	▒	■	□	0.25	1.004	4.557	5.665	10.095	16.772	22.996
3.	□	■	▒	□	□	0.25	17.834	43.669	51.006	86.091	113.111	120.049
4.	■	□	▒	□	□	0.25	38.927	64.879	72.905	106.778	135.889	152.651
5.	□	□	▒	■	■	0.50	3.929	6.842	7.556	12.298	19.115	27.983
6.	□	■	▒	□	■	0.50	22.003	52.932	46.579	90.029	119.882	135.002
7.	□	■	▒	■	□	0.50	25.991	55.014	49.082	93.332	122.229	138.909
8.	■	□	▒	□	■	0.50	40.02	66.698	74.117	108.183	137.710	154.119
9.	■	□	▒	■	□	0.50	44.384	70.192	78.251	112.672	141.008	158.725
10.	■	■	▒	□	□	0.50	62.193	89.002	96.621	130.298	159.027	186.791
11.	□	■	▒	■	■	0.75	37.971	61.995	71.003	105.391	134.441	160.219
12.	■	□	▒	■	■	0.75	46.929	72.982	80.388	114.901	143.412	169.590
13.	■	■	▒	□	■	0.75	65.129	91.109	99.041	133.923	162.721	189.993
14.	■	■	▒	■	□	0.75	68.905	94.795	102.003	136.729	165.11	192.10

A-Streptavidin | B-Protein | C-Biotin | D-APTES | E-Hydroprotein | F-Keratin

■ Presence of biomolecules □ Absence of biomolecules ▒ Oxide layer (SiO_2)

L_{CH1} = 14nm, L_{CH2} = 14nm, L_{CH3} = 2nm; $N_S = 10^{19}/cm^3$, $N_D = 10^{19}/cm^3$, $N_C = 10^{10}/cm^3$

biosensor with a prominent BioFET variant across different biomolecules [22–26]. The inset provides a visual representation of the threshold voltage sensitivity comparison between the proposed structure and four other variants. This comparison underscores that our biosensor offers superior sensitivity to biomolecules compared to most of the existing alternatives. Figure 7.19 shows the plot of gain transconductance frequency product (GTFP) for the proposed device at $N_S = N_D = 10^{19}/cm^3$ and $N_C = 10^{10}/cm^3$.

7.3.3 Machine Learning in Improving the BioFET Performance

Machine learning has significantly advanced the capabilities of BioFETs, by enhancing their sensitivity and specificity. A BioFET is a sophisticated biochemical sensor that operates on principles similar to those of a field effect transistor. This integration would permit a real-time investigation, which helps identify biological molecules like proteins, DNA, and pathogens. Training ML

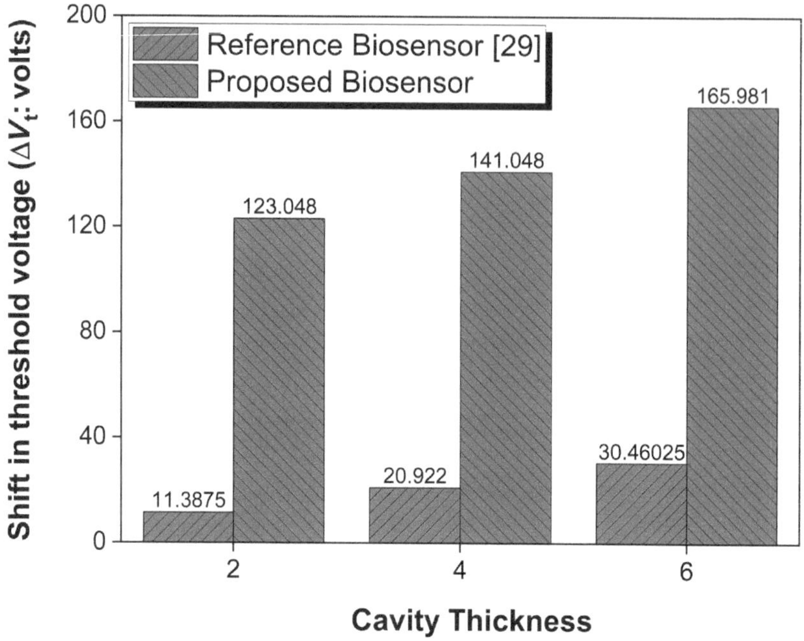

FIGURE 7.17 Comparison chart showing the shift in threshold voltage for APTES at different cavity thicknesses.

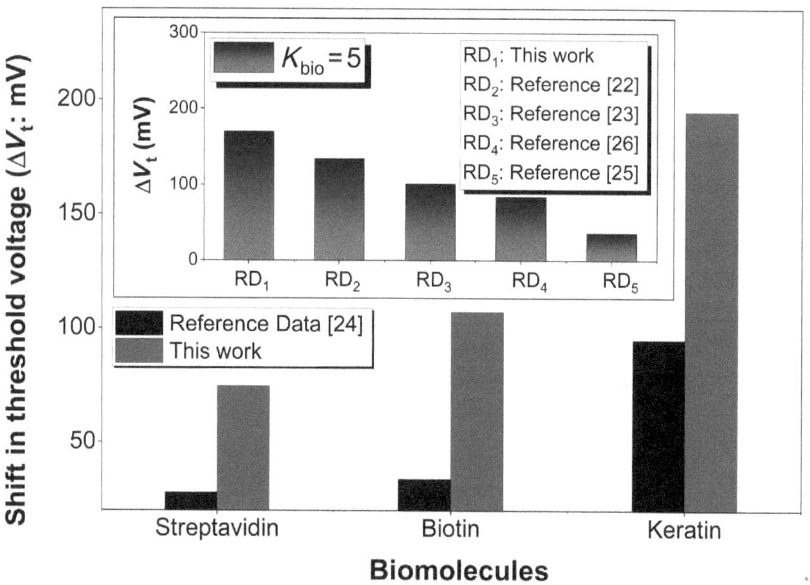

FIGURE 7.18 Comparison chart showing the shift in threshold voltage for different biomolecules. ([**INSET**] Shows the shift in threshold voltage in different variants of biosensor for $K_{bio} = 5$.)

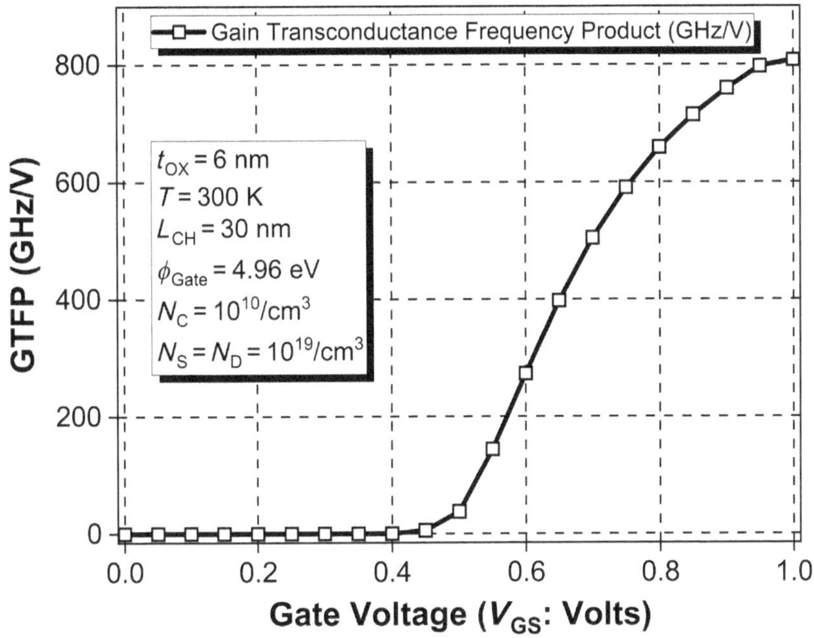

FIGURE 7.19 Gain transconductance frequency product plot.

models on such large datasets allows for the detection of subtle changes in sensor responses, leading to more effective detection and lower false favorable (detection) rates. Therefore, the advanced ML applications on BioFETs for medical diagnostics, environmental monitoring, and biotechnology shows an enhanced tool platform that can realize the rapid and accurate detection of numerous biological analytes [35].

Enabling the modification and tailoring of BioFETs for specific applications, using machine learning in combination with BioFETs, offers significant design benefits in biosensor development. Researchers can use sophisticated algorithms to model a range of conditions and geometries to target optimal sensor designs before they are fabricated. This speeds up the development process while reducing expensive trial-and-error methods. In addition, using historical data, machine learning can fine-tune sensor parameters to become more precise over time, so as new biological targets emerge, the sensor adapts with the changing times. The data that BioFETs produce is typically complex with multiple dimensions, which makes machine learning indispensable in interpreting such data. The influx of data that these sensors generate can be difficult to handle using traditional data analysis methods. On the other hand, ML models, such as neural networks, support vector machines (SVMs), and decision trees, are very good for identifying these correlations in big data collections. These approaches reveal more detailed details about how sensor responses relate to the biological interactions of the sensor surface for more accurate and reproducible results. An example can be found in medical diagnostics, where BioFETs with machine learning can differentiate between a normal and disease state, leading to identifying the early stages of a disease and more personal treatment strategies.

In extending to static analysis, machine learning also stirs the ideas of real-time monitoring and feedback systems. These conditions can change quickly in specific dynamic environments such as in vivo medical diagnostics or environmental monitoring. These scenarios need real-time data processing and a quick response as well. ML models continuously learn and adjust to new data streams, which is perfect for rapid analysis and taking action in real time. The real-time nature of BioFETs ensures the same level of accuracy and reliability, similar to vacuum-based benchtop instruments, and reliable performance in dynamic environments, which should be valuable in clinical and field situations. Small, user-friendly diagnostic devices also advance with the integration of machine learning and nanotechnology. Putting high-performance algorithms in small handheld devices allows us to do molecular biosensing in a nontraditional laboratory setting. Such portable BioFETs have valuable applications in point-of-care testing, remote health monitoring, and rapid field diagnostics, which will significantly facilitate the reach and convenience of biotechnology. The innate practice of machine learning underpinning these devices has enabled global access to laboratory-quality results on the fly and may well increase hallmark capabilities that could revolutionize health care practices worldwide. Finally, machine learning has much to implement in BioFETs to make them more functional, sensitive, and intelligent. This integration supports advanced data analysis, predictive modeling, real-time monitoring, and the creation of portable diagnostic tools. With the advent of new technologies and research, the intimate integration between machine learning and BioFETs will continue to foster algorithms in the biosensing diagnostic arena.

The introduction of machine learning on the road to developing BioFETs involves applying a few important considerations and techniques. These techniques encompass critical aspects:

1. **Data Preprocessing:** The first step is to prepare the raw data collected by hardware sensors for analysis. This entails data cleaning, standardizing, and improving data accuracy. Methods such as outlier detection, noise reduction, and normalization prepare the data. Principal component analysis (PCA) is one of the dimensionality reduction techniques, often leading to data dimension reduction while still preserving key information.
2. **Feature Extraction:** It is responsible for converting the unprocessed sensor data to something that can be used to feed into an ML model. For example, in a BioFET application, the features could be the current, voltage, or resistance changes due to biological interactions. Signal processing methods like Fourier transform and wavelet transform extract frequency domain features of a time-domain sensor signal. Convolutional neural networks (CNNs) are an example of advanced methods to automate the extraction of complex features and are especially well suited for discerning intricate patterns of biological interactions.

3. **Choosing the Right Model:** Accurate data analysis depends on the ML technique used. SVMs, neural networks (including CNNs and deep neural networks or DNNs), random forests, and *K*-nearest neighbors (KNN) are some of the models selected considering the nature of the data and the type of analysis required. The properties of each model type make them advantageous in tackling different forms and complexities of BioFET data.
4. **Adaptation in Real Time:** With real-time adaptation, models can be updated periodically when new data is available to keep models accurate even with environmental changes. For this purpose, more advanced methods like online learning and reinforcement learning, among others, are employed. In online learning, when new data comes in, learning algorithms increase or decrease your model parameters. Thus, this type of learning is very appropriate for applications that require high real-time response. Employment of BioFETs in combination with reinforcement learning allows BioFETs to learn the best response on demand from real-time feedback, which is a must for any sensor designed to operate in the inherently complicated biological environment and be capable of interacting with a diverse set of targets and conditions.
5. **Ensemble Methods:** In ensembling, instead of making single predictions, multiple models work simultaneously, and the expected outcome enhances general performance and robustness. Another group of approaches includes bagging, boosting, and stacking, which combines predictions and reduces the effect of overfitting that could occur during the implementation of BioFET.
6. **Transfer Learning:** Fine-tuning enhances the BioFET model creation process by using pretrained models from related tasks, which helps to increase the accuracy of predicting priorities when the dataset is small. It is seen that BioFETs can not only help advance function-based models but also help with faster deployment when it comes to biosensing and diagnostics.
7. **Predictive Maintenance and Calibration:** The data collected enables automated monitoring for the optimal condition, calibration, and service schedule of BioFET devices using the remaining data analysis from machine learning. Analytical models are developed to predict other maintenance requirements and calibration frequencies, enabling the sensor's precise control and lifelong performance.

ML models are used in developing BioFETs to boost the sensitivity, specificity, and flexibility of sensors. Before applying an accurate real-time adaptive model selection algorithm, combined with advanced feature extraction, using enhanced preprocessing techniques, ensemble methods, and actual transfer learning, to the biosensing and diagnostics of BioFET sensors, it is crucial to ascertain their reliability as tools in medical, environmental, and biotechnological applications.

Data Analysis Based on ML: PCA-based analysis, SVM-based analysis, and artificial neural network (ANN) based analysis are three distinct ML approaches widely employed in various fields, including BioFET sensors.

1. **PCA-Based Analysis:** PCA is a statistical technique used primarily for dimensionality reduction and feature extraction:
   - **Purpose:** PCA identifies essential patterns or components within complex, high-dimensional sensor data.
   - **Process:** It transforms original variables into a smaller set of orthogonal components (principal components) that retain the variability present in the data.
   - **Application:** In BioFETs, PCA simplifies sensor responses while preserving crucial information about biological interactions. It aids in visualizing data patterns and identifying influential/significant variables affecting sensor outputs.
2. **SVM-Based Analysis:** SVM are robust, influential models for classification and regression, which are types of supervised learning models:
   - **Purpose:** The findings indicate that SVMs help classify the biological targets from the BioFET sensor dataset.
   - **Process:** Maximum margin classification machines delineate separate classes in the high-dimensional space by constructing optimal hyperplanes.
   - **Application:** In BioFETs, SVMs can classify the sensor responses of either biomolecules or disease states. An advantage of using polynomial-based methods when dealing with interactions is that they are not sensitive to overfitting, are good at managing nonlinear relationships, and are ideal when dealing with biological interactions.
3. **ANN-Based Analysis:** ANNs, including DNNs, are mathematical models inspired by biological neural networks; they excel at efficiently learning from complex patterns.
   - **Purpose:** ANNs can recognize patterns within large datasets where the patterns are complicated.
   - **Process:** They consist of multiple levels of neurons connected to each other and derive hierarchical dimensions of data.
   - **Application:** ANNs in BioFETs process sensor information to predict interaction with biological systems or recognize complex patterns. CNNs within ANNs are particularly well suited to handling spatial sensor outputs, while recurrent neural networks (RNNs) perform well with more sequential data, such as time-series responses. It is a class of models that needs much computational power but can achieve very accurate and adaptable solutions for various problems.

PCA is an analytical solution that is concerned with dimensionality reduction and the primary examination of the dataset at hand without necessarily using training or testing data. CNNs and RNNs are types of ANNs used to address intricate patterns and are computationally intensive. These methods are beneficial and can be integrated when used to analyze data from BioFETs. PCA is best utilized in its implementation to remove noise and develop features to feed an SVM or ANN. SVM tends to work better in categorizing datasets, while ANNs help fine-tune or deal with complex data relationships. In conclusion, PCA, SVM, and ANN hold great power for data analysis of BioFET arrays, and every method has its distinct advantages in terms of dimensionality reduction, classification, and pattern analysis. These methods complement each other in improving the accuracy, speed, and analysis associated with BioFET applications in the biomedical and biotechnological sectors.

BioFET Design Based on ML: More recently, the design of sensors, particularly BioFETs, has benefited from using ML algorithms in the process. These applications revolutionize sensor development by enhancing operating efficiency, upgrading materials, shortening the design cycle time, and selecting a suitable receptor.

1. **Enhancing BioFET Performance:** Machine learning is essential when it comes to improving the performance of BioFETs for the following reasons:
   - **Parameter Optimization:** Machine learning is vital for enhancing BioFET performance by optimizing various parameters for better sensitivity, selectivity, and overall efficiency [62].
   - **Material Design:** ML algorithm applications are critical in augmenting the 2D material characteristics that are singularly employed in BioFETs, including semiconducting transition metal dichalcogenides (TMDCs). ML is found to enhance the sensing system's reliability and performance by predicting parameters such as bandgaps and facilitating the optimization of defect structures [63].
2. **Accelerated Design Processes:** These are due to key ML features—that is, the ability to support bioelectronics design and development through enhancing the application for rapid prototyping and iterative designing of BioFETs:
   - **Heterostructure Optimization:** There are helpful methods, such as Gaussian process regression and Bayesian optimization, to initiate the design of complicated heterostructures in BioFETs. These techniques anticipate things such as electrical and optical characteristics and help to identify the best formula for sensor materials to improve the functionality of sensors [64].
   - **Organic FET Enhancement:** Other learning algorithms, such as random forest and gradient boosting, significantly improve electron flow in organic FETs. The ML models link electronic properties to sensor performance, thus enhancing the chances of accurate detection and even the level of sensitivity [65].
3. **Advanced Receptor Selection:** Machine learning transforms the picking (bioreceptor layer) and tuning of receptors to enhance the possibility of BioFET sensing.
   - **Receptor Engineering:** ML-driven approaches predict and optimize receptors such as enzymes and aptamers based on data patterns. This method surpasses traditional approaches by identifying novel receptors with high specificity and efficiently detecting target molecules [66].
4. **Integrating ML for BioFET Advancements:** Altogether, the incorporation of machine learning into BioFET design is deemed a revolutionary strategy, allowing the fabrication of sensors from essential elements consisting

of biologically related and electronic components in a highly accurate and efficient way in terms of sensitivity, specificity, and versatility. ML methods in optimizing parameters, construction of materials, and rapid prototyping for micro-receptor selection in BioFETs have placed them at the strategic center of health diagnostics, environmental monitoring, pollution tracking, and disease detection. These developments clearly highlight the role of ML in the development of sensor technologies and the capability of utilizing it to solve challenging issues in various domains.

7.3.4 Discussion

Doping significantly influences a BioFET's sensitivity, with higher percentage sensitivity observed at lower doping levels for both threshold voltage and subthreshold slope. Sensitivity further improves with higher dielectric constant biomolecules, and negatively charged biomolecules exhibit enhanced sensitivity due to their properties (negatively charged behavior). Biomolecule location within partially filled cavities, especially near the source, significantly impacts threshold voltage sensitivity, overshadowing the fill-in factor effect. The sensitivity was evaluated using various metrics, including threshold voltage, subthreshold slope, I_{ON}/I_{OFF} ratio, and drain current. The novelty of this work lies in two main aspects: first, enhancing sensitivity through optimizing doping and other parameters, and second, comprehensively exploring partially filled cavities at different fill-in factors. The use of various biomolecules further validates research findings.

Optimizing doping and bias settings is critical in the prefabrication phase of BioFETs. These findings underscore the efficacy of this biosensor in detecting essential biomolecules crucial for diverse biomedical and clinical applications. Additionally, it holds promise for rapidly detecting various other medically significant biomolecules, thereby aiding in the early diagnosis of acute and chronic diseases. This work demonstrates the potential of BioFETs as robust and sensitive sensors, highlighting their broader applicability within the biomedical industry, which warrants further exploration.

The proposed BioFET also offers a cost-effective and safe alternative for remotely detecting hazardous chemicals, addressing potential health risks. Nano-sized BioFET-based wearable devices hold great promise for detecting harmful gases or radiation in environments like reactive chambers and monitoring health conditions in personalized and regular checkups for patients or elderly individuals. Additionally, the research delves into the critical role and increasing necessity of integrating machine learning into the BioFET design process. This discussion highlights how ML algorithms can enhance the sensitivity, accuracy, and overall performance of BioFETs by analyzing complex biochemical interactions and optimizing sensor parameters.

7.4 CONCLUSION

This chapter investigates the sensitivity analysis of a cylindrical BioFET, focusing mainly on the influence of doping levels. By exploring a range of charged and neutral biomolecules, the study enhances the realism and relevance of its findings.

It underscores that threshold voltage sensitivity outstrips that of the subthreshold slope, with the sensitivity of the former metric exceeding 400%, which is achievable through meticulously optimized doping levels, highlighting the device's efficacy for biosensing applications. In the symmetric doping case for source and drain, peak fractional sensitivities for threshold voltage and subthreshold slope reach up to 4.53 and 0.44, respectively. It was found in the study that high percentage sensitivity was obtained at low doping levels, with a trade-off with conductivity. The research also scrutinizes the impact of fill-in factors, revealing that the precise placement of biomolecules near the source significantly affects sensitivity more than the actual fill-in factor itself. Considering V_t, as a sensing metric, a high percentage sensitivity is achieved at low doping levels, while a high relative change sensitivity is achieved at high doping levels. Overall, this study emphasizes the pivotal roles of doping levels, bias conditions, and biomolecule placement in determining sensitivity in a cylindrical BioFET. The study also discussed the role of machine learning in improving the performance and design of BioFETs through a synoptic discussion.

ACKNOWLEDGMENTS

The authors are grateful to the director, Maharaja Agrasen Institute of Technology, Delhi, for providing the facilities to carry out this research work.

REFERENCES

1. K. Upasana, R. Narang, M. Saxena, and M. Gupta, "Modeling and TCAD assessment for gate material and gate dielectric engineered TFET architectures: Circuit-level investigation for digital applications," *IEEE Trans. Electron Devices*, vol. 62, no. 10, pp. 3348–3356, 2015.
2. A. Sirohi, C. Sahu, and J. Singh, "Analog/RF performance investigation of dopingless FET for ultra-low power applications," *IEEE Access*, vol. 7, pp. 141810–141816, 2019.
3. J. Park, H. Hiep, A. Woubit, and M. Kim, "Applications of field effect transistor (FET) – type biosensors," *Appl. Sci. Converg. Technol.*, vol. 23, no. 2, pp. 61–71, 2014.
4. A. Das, B. K. Kanaujia, V. Nath, S. Rewari, and R. S. Gupta, "Impact of reverse gate oxide stacking on gate all around tunnel FET for high frequency analog and RF applications," in *2020 IEEE 17th India Council International Conference, INDICON 2020*, pp. 1–6, 2020.
5. R. Gautam, M. Saxena, R. S. Gupta, and M. Gupta, "Gate all around nanowire MOSFET with catalytic metal gate for gas sensing applications," *IEEE Trans. Nanotechnol.*, vol. 12, no. 6, pp. 939–944, 2013.
6. T. Bungon *et al.*, "Graphene FET sensors for Alzheimer's disease protein biomarker clusterin detection," *Front. Mol. Biosci.*, vol. 60, no. 1, pp. 1–13, 2021.
7. S. Cheng, S. Hideshima, S. Kuroiwa, T. Nakanishi, and T. Osaka, "Label-free detection of tumor markers using field effect transistor (FET)-based biosensors for lung cancer diagnosis," *Sensors Actuators, B Chem.*, vol. 212, pp. 329–334, 2015, http://dx.doi.org/10.1016/j.snb.2015.02.038.
8. Y.-C. Syu, W.-E. Hsu, and C.-T. Lin, "Review – field-effect transistor biosensing: Devices and clinical applications," *ECS J. Solid State Sci. Technol.*, vol. 7, no. 7, pp. Q3196–Q3207, 2018.
9. M. Kaisti, "Detection principles of biological and chemical FET sensors," *Biosens. Bioelectron.*, vol. 98, pp. 437–448, 2017.

10. A. Das, S. Rewari, B. K. Kanaujia, and R. S. Gupta, "Recent technological advancement in surrounding gate MOSFET for biosensing applications – a synoptic study," *Silicon*, vol. 14, no. 10, pp. 5133–5143, 2022.
11. V. K. Tamboli *et al.*, "Hybrid synthetic receptors on MOSFET devices for detection of prostate specific antigen in human plasma," *Anal. Chem.*, vol. 88, no. 23, pp. 11486–11490, 2016.
12. A. Das, A. Bhardwaj, K. Das, S. Yadav, A. Kaul, P. Goyal, S. Sharma, S. Rewari, B. K. Kanaujia, and R. S. Gupta, "Numerical simulation on sensitivity modulation in nanowire MOSFEB detector for biomedical applications," *Indian J. Pure Appl. Phys.*, vol. 63, pp. 281–300, 2025, https://doi.org/10.56042/ijpap.v63i4.14388.
13. K. Kaushik, and D. Rawtani, "Sensor-based techniques for detection of COVID-19," in *COVID-19 in the Environment*, Elsevier, pp. 95–114, 2022.
14. T. Cheung, P. K. N. Yu, and M. J. Butson, "Low-dose measurement with a MOSFET in high-energy radiotherapy applications," *Radiat. Meas.*, vol. 39, no. 1, pp. 91–94, 2005.
15. R. Consorti *et al.*, "In vivo dosimetry with MOSFETs: Dosimetric characterization and first clinical results in intraoperative radiotherapy," *Int. J. Radiat. Oncol. Biol. Phys.*, vol. 63, no. 3, pp. 952–960, 2005.
16. J. R. Grenier, S. H. Jayaram, M. Kazerani, H. Wang, and M. W. Griffiths, "MOSFET-based pulse power supply for bacterial transformation," *IEEE Trans. Ind. Appl.*, vol. 44, no. 1, pp. 25–31, 2008.
17. T. Morf *et al.*, "Wide bandwidth room-temperature THz imaging array based on antenna- coupled MOSFET bolometer," *Sensors Actuators, A Phys.*, vol. 215, no. 15, pp. 96–104, 2014.
18. C. Jianrong, M. Yuqing, H. Nongyue, W. Xiaohua, and L. Sijiao, "Nanotechnology and biosensors," *Biotechnol. Adv.*, vol. 22, no. 7, pp. 505–518, 2004.
19. A. Das, S. Rewari, B. K. Kanaujia, S. S. Deswal, and R. S. Gupta, "Numerical modeling of a dielectric modulated surrounding-triple-gate germanium-source MOSFET (DM-STGGS-MOSFET)-based biosensor," *J. Comput. Electron.*, vol. 22, no. 2, pp. 742–759, 2023, https://doi.org/10.1007/s10825-023-02008-w.
20. A. Chakraborty, and A. Sarkar, "Analytical modeling and sensitivity analysis of dielectric modulated junctionless gate stack surrounding gate MOSFET (JLGSSRG) for application as biosensor," *J. Comput. Electron.*, vol. 16, no. 3, pp. 556–567, 2017.
21. A. Das, S. Rewari, B. K. Kanaujia, S. S. Deswal, and R. S. Gupta, "Ge/Si interfaced label free nanowire BIOFET for biomolecules detection – analytical analysis," *Microelectronics J.*, vol. 138, p. 105832, 2023, https://doi.org/10.1016/j.mejo.2023.105832.
22. K. N. Singh, and P. K. Dutta, "Analytical modeling of a high-k underlap dielectric and charge modulated silicon-on-nothing FET based biosensor," *J. Comput. Electron.*, vol. 19, no. 3, pp. 1126–1135, 2020.
23. S. Ajay, R. Narang, M. Gupta, and M. Saxena, "Investigation of dielectric modulated double gate junctionless MOSFET for detection of biomolecules," in *2013 Annual IEEE India Conference, INDICON 2013*, pp. 1–6, 2013.
24. A. Kumar, N. Gupta, M. M. Tripathi, and R. Chaujar, "Analysis of structural parameters on sensitivity of black phosphorus junctionless recessed channel MOSFET for biosensing application," *Microsyst. Technol.*, vol. 26, no. 7, pp. 2227–2233, 2020.
25. S. Ajay, R. Narang, M. Saxena, and M. Gupta, "Investigation of dielectric modulated (DM) double gate (DG) junctionless MOSFETs for application as a biosensors," *Superlattices Microstruct.*, vol. 85, pp. 557–572, 2015.
26. S. Sharma, V. Nath, S. S. Deswal, and R. S. Gupta, "Analytical modelling and sensitivity analysis of Gallium nitride-gate material and, dielectric engineered-Schottky nanowire FET (GaN-GME-DE-SNW-FET) based label-free biosensor," *Microelectronics J.*, vol. 129, p. 105599, 2022, https://doi.org/10.1016/j.mejo.2022.105599.

27. Z. Ahangari, "Performance assessment of dual material gate dielectric modulated nanowire junctionless MOSFET for ultrasensitive detection of biomolecules," *RSC Adv.*, vol. 6, no. 92, pp. 89185–89191, 2016.
28. C. Li, F. Liu, R. Han, and Y. Zhuang, "A vertically stacked nanosheet gate all around FET for biosensing application," *IEEE Access*, vol. 9, pp. 63602–63610, 2021.
29. Y. Pratap, M. Kumar, S. Kabra, S. Haldar, R. S. Gupta, and M. Gupta, "Analytical modeling of gate all around junctionless transistor based biosensors for detection of neutral biomolecule species," *J. Comput. Electron.*, vol. 17, no. 1, pp. 288–296, 2018.
30. X. P. A. Gao, G. Zheng, and C. M. Lieber, "Subthreshold regime has the optimal sensitivity for nanowire FET biosensors," *Nano Lett.*, vol. 10, no. 2, pp. 547–552, 2010.
31. A. AssarehPour, and Farhad Shahhoseini, "Gas sensor based on Silicon nanotube MOSFET," in *5th International Congress on Nanoscience & Nanotechnology*, pp. 1–2, 2014.
32. T. V. A. Dam, D. Pijanowska, W. Olthuis, and P. Bergveld, "Highly sensitive glucose sensor based on work function changes measured by an EMOSFET," *Analyst*, vol. 128, no. 8, pp. 1062–1066, 2003.
33. A. Das, S. Rewari, B. K. Kanaujia, S. S. Deswal, and R. S. Gupta, "Analytical investigation of a triple surrounding gate germanium source metal – oxide – semiconductor field-effect transistor with step graded channel for biosensing applications," *Int. J. Numer. Model. Electron. Networks, Devices Fields*, vol. 36, no. 6, p. e3106, 2023, https://doi.org/10.1002/jnm.3106.
34. S. S. Mohanty, S. Mishra, M. Mohapatra, and G. P. Mishra, "Impact of bio-target location and their fill-in factor on the sensitivity of hetero channel double gate MOSFET label-free biosensor," *Adv. Nat. Sci. Nanosci. Nanotechnol.*, vol. 12, no. 2, p. 025012, 2021, https://doi.org/10.1109/DevIC50843.2021.9455937.
35. X. Xhafa, A. L. I. D. O. Gu, and M. B. Yelten, "Machine learning-based modeling of hot carrier injection in 40 nm CMOS transistors," *J. Electron Devices Soc.*, vol. 12, pp. 281–288, 2024.
36. H. Amrouch, V. M. van Santen, J. Diaz-Fortuny, and F. Klemme, "Machine learning unleashes aging and self-heating effects: From transistors to full processor," in *2024 IEEE International Reliability Physics Symposium (IRPS)*, pp. 1–8, 2024, https://doi.org/10.1109/IRPS48228.2024.10529386.
37. S. S. Adel, and K. C. Smith, *Microelectronics Circuit*, 8th ed. Oxford, 2020.
38. A. Das, S. Rewari, B. K. Kanaujia, S. S. Deswal, and R. S. Gupta, "Physics based numerical model of a nanoscale dielectric modulated step graded germanium source biotube FET sensor: Modelling and simulation," *Phys. Scr.*, vol. 98, no. 11, p. 115013, 2023, https://doi.org/10.1088/1402-4896/acf4c9.
39. G. Jana, D. Sen, P. Debnath, and M. Chanda, "Power and delay analysis of dielectric modulated dual cavity junctionless double gate field effect transistor based label-free biosensor," *Comput. Electr. Eng.*, vol. 99, p. 107828, April 2022.
40. E. Rahman, A. Shadman, and Q. D. M. Khosru, "Effect of biomolecule position and fill in factor on sensitivity of a dielectric modulated double gate junctionless MOSFET biosensor," *Sens. Bio-Sensing Res.*, vol. 13, pp. 49–54, 2017.
41. Y. Zhang, K. Han, and J. Li, "A simulation study of a gate all around nanowire transistor with a core-insulator," *Micromachines*, vol. 11, no. 2, pp. 1–12, 2020.
42. A. Ganesh, K. Goel, J. S. Mayall, and S. Rewari, "Subthreshold analytical model of Asymmetric Gate Stack Triple Metal Gate all Around MOSFET (AGSTMGAAFET) for improved analog applications," *Silicon*, vol. 14, pp. 4063–4073, 2021.
43. S. Ye, K. Yamabe, and T. Endoh, "Ultimate vertical gate all around metal oxide semiconductor field effect transistor and its three dimensional integrated circuits," *Mater. Sci. Semicond. Process.*, vol. 134, p. 106046, 2021.
44. S. M. Sze, *VLSI Technology*, 2nd ed. McGraw Hill Education, 2017.

45. A. Das, S. Rewari, B. K. Kanaujia, S. S. Deswal, and R. S. Gupta, "Analytical characterization of a label free Si/InAs hetero-interfaced cylindrical BioFETD for biosensing applications," *Micro Nanostruct.*, vol. 204, p. 208152, 2025, https://doi.org/10.1016/j.micrna.2025.208152.
46. S. Yadav, A. Das, and S. Rewari, "Dielectrically-modulated GANFET biosensor for label-free detection of DNA and Avian influenza virus: Proposal and modeling," *ECS J. Solid State Sci. Technol.*, vol. 13, no. 4, p. 047001, 2024, https://doi.org/10.1149/2162-8777/ad3364.
47. Y. Pratap, M. Kumar, S. Kabra, S. Haldar, R. S. Gupta, and M. Gupta, "Analytical modeling of gate-all-around junctionless transistor based biosensors for detection of neutral biomolecule species," *J. Comput. Electron.*, vol. 17, no. 1, pp. 288–296, 2018.
48. R. G. Wilson, "Vacuum thermionic work functions of polycrystalline Nb, Mo, Ta, W, Re, Os, and Ir," *J. Appl. Phys.*, vol. 37, no. 8, pp. 3170–3172, 1966.
49. H. Kawano, T. Takahashi, Y. Tagashira, H. Mine, and M. Moriyama, "Work function of refractory metals and its dependence upon working conditions," *Appl. Surf. Sci.*, vol. 146, no. 1, pp. 105–108, 1999.
50. A. Goel, S. Rewari, S. Verma, and R. S. Gupta, "Dielectric modulated triple metal gate all around MOSFET (TMGAA) for DNA biomolecule detection," in *Proceedings of International Conference on 2018 IEEE Electron Device Kolkata Conference, EDKCON 2018*, pp. 337–340, 2018.
51. S. Busse, V. Scheumann, B. Menges, and S. Mittler, "Sensitivity studies for specific binding reactions using the biotin/streptavidin system by evanescent optical methods," *Biosens. Bioelectron.*, vol. 17, no. 8, pp. 704–710, 2002.
52. A. Goel, S. Rewari, S. Verma, S. S. Deswal, and R. S. Gupta, "Dielectric modulated junctionless biotube FET (DM-JL-BT-FET) biosensor," *IEEE Sens. J.*, vol. 21, no. 15, pp. 16731–16743, 2021.
53. K. N. Singh, and P. K. Dutta, "Analytical modeling of underlap graded channel field effect transistor as a label free biosensor," *Superlattices Microstruct.*, vol. 155, p. 106897, July 2021.
54. A. Lodhi, C. Rajan, A. Kumar, B. Dip, P. Samajdar, and D. Soni, "Sensitivity and sensing speed analysis of extended nano cavity and source over electrode in Si/SiGe based TFET biosensor," *Appl. Phys. A*, vol. 126, no. 11, pp. 1–8, 2020.
55. E. Rosini, P. D'Antona, and L. Pollegioni, "Biosensors for d-amino acids: Detection methods and applications," *Int. J. Mol. Sci.*, vol. 21, no. 13, p. 4574, 2020.
56. A. Das, B. K. Kanaujia, S. S. Deswal, S. Rewari, and R. S. Gupta, "Doping induced threshold voltage and ION/IOFF ratio modulation in surrounding gate MOSFET for analog applications," in *2022 IEEE International Conference of Electron Devices Society Kolkata Chapter (EDKCON)*, Kolkata, India, pp. 1–6, 2022, https://doi.org/10.1109/EDKCON56221.2022.10032940.
57. A. Das, S. Rewari, B. K. Kanaujia, S. S. Deswal, and R. S. Gupta, "Analytical modeling and doping optimization for enhanced analog performance in a Ge/Si interfaced nanowire MOSFET," *Phys. Scr.*, vol. 98, no. 7, p. 74005, 2023, https://doi.org/10.1088/1402-4896/acde16.
58. S.-M. Kang, and L. Yusuf, *CMOS Digital Integrated Circuits*, 3rd ed., McGraw Hill, 2003.
59. D. M. Ã, J. Song, and O. Kim, "Effect of source/drain doping gradient on threshold voltage variation in double gate fin-field effect transistors as determined by discrete random doping," *Jpn. J. Appl. Phys.*, vol. 104301, no. 49, p. 4, 2010.
60. A. Bhardwaj, A. Das, U. Sharma, A. Gupta, and S. Roy, "Temperature induced analog performance modulation of high-k vertical nanowire tunnel FET," *Semiconductors*, vol. 59, no. 4, pp. 382–390, 2025, https://doi.org/10.1134/S1063782625600020.
61. M. Kumari, N. K. Singh, M. Sahoo, and H. Rahaman, "Work function optimization for enhancement of sensitivity of dual material (DM), double gate (DG), junctionless MOSFET based biosensor," *Appl. Phys. A Mater. Sci. Process.*, vol. 127, no. 2, pp. 1–8, 2021.

62. B. Cao *et al.*, "How to optimize materials and devices via design of experiments and machine learning: demonstration using organic photovoltaics," *ACS Nano*, vol. 12, pp. 7434–7444, 2018.
63. N. C. Frey, D. Akinwande, D. Jariwala, and V. B. Shenoy, "Machine learning-enabled design of point defects in 2D materials for quantum and neuromorphic information processing," *ACS Nano*, vol. 14, no. 10, pp. 13406–13417, 2020.
64. L. Bassman *et al.*, "Active learning for accelerated design of layered materials," *npj Comput. Mater.*, vol. 4, no. 1, p. 71, 2018, http://dx.doi.org/10.1038/s41524-018-0129-0.
65. M. Lee, "Machine learning for understanding the relationship between the charge transport mobility and electronic energy levels for n-type organic field-effect transistors," *Adv. Electron. Mater.*, vol. 5, no. 12, p. 1900573, 2019.
66. J. Song *et al.*, "A sequential multidimensional analysis algorithm for aptamer identification based on structure analysis and machine learning," *Anal. Chem.*, vol. 92, pp. 3307–3314, 2020.

8 Semiconductor Materials for EV and Renewable Energy

Manjulata Badi and Ila Rai

8.1 INTRODUCTION

The advancements in pressure semiconductor generation and its use in electric and hybrid vehicles (HEV/EV) are discussed in this chapter. A review of the latest automotive pressure semiconductor devices is included, along with an analysis of the theories and present state of the pressure hardware required for EVs and HEVs. The Si-based insulated gate bipolar transistor (IGBT), the SiC-based metal oxide semiconductor field effect transistor (MOSFET), the freewheeling diode (FRD), and the Schottky obstacle diode are a few of the most recent innovations in the automotive semiconductor sector. Reduced heat opposition planning, high-degree bundling, and low inductance in high-strength thickness bundling are the three characteristics that define the advancements in automotive semiconductor bundling technologies.

We talk about the latest, higher-power semiconductors for EVs and HEVs, their demanding environments, and new developments in this field. For Si devices, the main challenges are strength, thickness, performance, stability, and bundling; for SiC devices, the main challenges are low inductances and high temperatures. The long-term improvement trend is analyzed from four angles: planar bundling, novel bundling innovation and material, and the SiC MOSFET. Features like flip-around conduction-IGBT, recessed producer channel, and amazing IGBT are all part of the next-generation Si-based IGBT for HEV/EV. Sustainable growth requires the development of affordable, widely available, and environmentally friendly renewable energy technologies. Technology for (photo)electrochemical energy conversion and storage is a crucial element. Therefore, to really attain to the practical use of (photo) electrochemical energy technology, realistic design, synthesis, and modification of nanostructured catalytic materials are needed. Ion beam technology is one practical and versatile physical modification method. By controlling the energy, species, and fluence of ions that are implanted, it is possible to modify the surface, interface, and thin film of various catalytic material types. The ion beam approach has several benefits, including compulsivity of element doping and great controllability, precision, and reproducibility [1]. The recent advances and trends in HEV/EV-oriented power semiconductors is depicted in Figure 8.1.

The global transaction data and market gauge for HEV/EV converters are shown in Figure 8.1. The four main improvements to contemporary power digital frameworks

 DOI: 10.1201/9781003508304-8

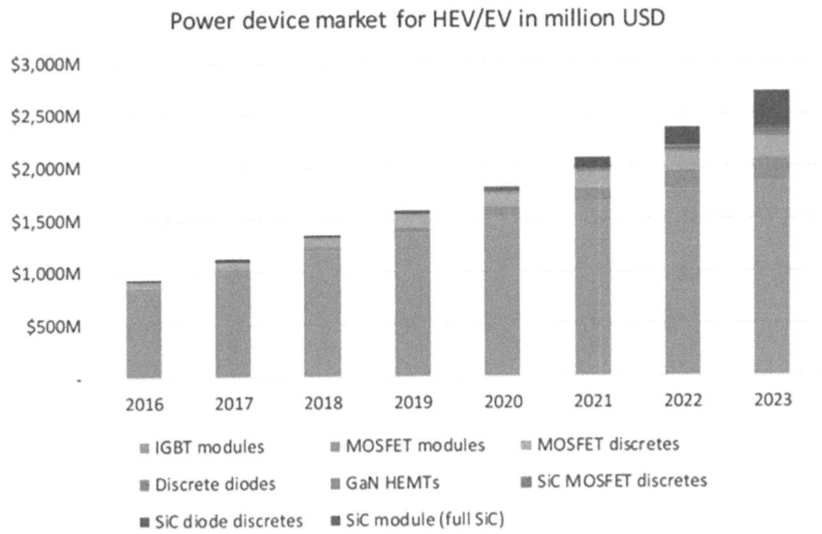

FIGURE 8.1 Recent advances and trend of HEV/EV-oriented power semiconductors: An overview.

Source: Liu, G., Li, K., Wang, Y., Luo, H., & Luo, H. (2020). Recent advances and trend of HEV/EV-oriented power semiconductors – an overview. ***IET Power Electronics, 13*****(3), 394–404, https://doi.org/10.1049/iet-pel.2019.0401.**

in hybrid and electric cars are the onboard charger, the DC dollar converter for assistance power resources, the DC assist converter in the high-voltage transport bar, and the inverter component for engine pressure. According to research, the industry will grow over the next five years and surpass ten billion dollars with high-power HEV/EV converters have so far been created using a range of contemporary power semiconductor types, including IGBTs, metal oxide semiconductor subject-impact transistors (MOSFET), and extensive-band-hole (WBG) devices, including silicon carbide (SiC) and gallium nitride (GaN) devices. Figure 8.2 shows the recent advances and trends in HEV/EV.

The agreements and anticipated pressure device market scenario for HEV/EV are shown in Figure 8.2. With over 80% of the present HEV/EV electrical module market, it has been determined that IGBT is the most extensively utilized module in the business. Moreover, the whole SiC module is predicted to just narrowly lose out on victory (by around 13%) over a five-year period. In strength electronic frameworks, the strength module could be the weakest element. According to an enterprise investigation using data from over 80 agencies, temperature element activation accounts for 55% of pressure electronic system failures at present, with power module failures accounting for around 42% of them. Moreover, for every degree that the temperature rises, the intersection's misery rate climbs by 10°C. Regarding the power semiconductors in the HEV/EV framework [2]. The unpredictable work stresses of the current era create ever-stricter requirements on the power electronic modules of electric vehicle systems due to unpredictable weather

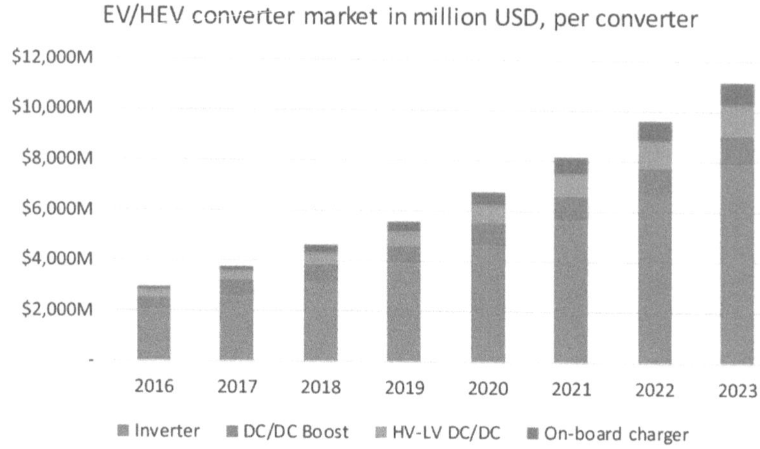

FIGURE 8.2 Recent advances and trend of HEV/EV-convertor market: An overview.

and road conditions. At this point, it is economically feasible to operate Si-based IGBTs and SiC-based full MOSFETs at 175°C [3]. The highest potential temperature for recreational use of SiC sheet is 500°C, which is significantly higher than the maximum temperature for modern Si material. SiC electrical devices, however, seem to make more sense for packages that require high temperature reliability and severe climates. Because of its high temperature (>500°C), high electron velocity (321 cm/s), and high breakdown electric power field (2.2 × 106 V/cm) characteristics, the SiC-based device is a potential candidate for HEV/EV converter projects. SiC-based MOSFET modules, as delegates in WBG devices, are gradually changing the conventional utility domain names that concern Si-based MOSFETs and IGBTs [4]. Nevertheless, in terms of material value and sophisticated handling capabilities, SiC semiconductors have surely lagged behind such state-of-the-art Si-based devices. However, the final demise of SiC is not well explained by the same old wire-bolstered package architectures. Based on this idea, a myriad of emerging technologies have emerged in an effort to address these difficulties from the perspective of a modern, cutting-edge package offer. As the engine behind the use of electric power, hybrid electric powered motors (HEVs) with an energy semiconductor base are undergoing a constant change and are at a turning point in their history. The latest force module changes for stylish exhibitions by the public are a consequence of breakthroughs in semiconductor and DB technology. To find new development opportunities for future exploratory research and HEV/EV firms, this study looks at the developments in new automotive strength semiconductors and the problems and methodology when it comes to developing more robust electronic frameworks in HEV/EV.

The increasing realization of the meaning of state-of-the-art solar power has led to strengthened research in the field of today's sun-oriented energy collection. In contrast to silicon (Si), electricity instruments are becoming a major part of power hardware innovation. The performance of the photovoltaic (PV) inverter, a crucial part of the solar-powered energy trading system, depends on the clever design of state-of-the-art pressure hardware and is depicted in Figure 8.3, with the silicon carbide substrate market in units and dollars for PV inverters. To maximize the power extracted from daylight-based chargers, the strength anguish inside the electricity change framework should be controlled by suitable determination of contemporary semiconductor devices, consequently limiting the quantity of today's electricity digital elements. The need to lower the overall switch mass and extent has prompted the creation of the newest extreme-strength in Zeng and others [5]. Figure 8.3 summarizes the PV inverters in the SiC substrate market.

The relevance of fresh silicon carbide semiconductor innovation is demonstrated by the cutting-edge SiC function in the aforementioned dwellings. Some of the constraints of today's innovation are higher system costs, the need for extreme-temperature bundling procedures, and newer inventions. Subsequent investigations will concentrate on techniques to reduce manufacturing expenses, resolve packaging problems, and surmount obstacles to enhance the dependability and visual appeal of contemporary SiC devices. The primary focus of this chapter is the process of creating potent devices based on SiC semiconductors within the context of PV energy transformation [6]. Furthermore, experiments with different PV inverter structures indicate that using novel SiC energy semiconductor devices within a PV strength framework might assist in removing some problems that might be presently generating material barriers to contemporary silicon. Future developments in solar technology have made it possible to produce more usable electricity. The power that is extracted from the sun and transformed into thermal or electrical energy is known as solar-oriented strength. It's an amazing resource that is widely available everywhere in the world. Application-scale primarily daylight-based energy flowers (flowers that generate a significant amount of new energy that the

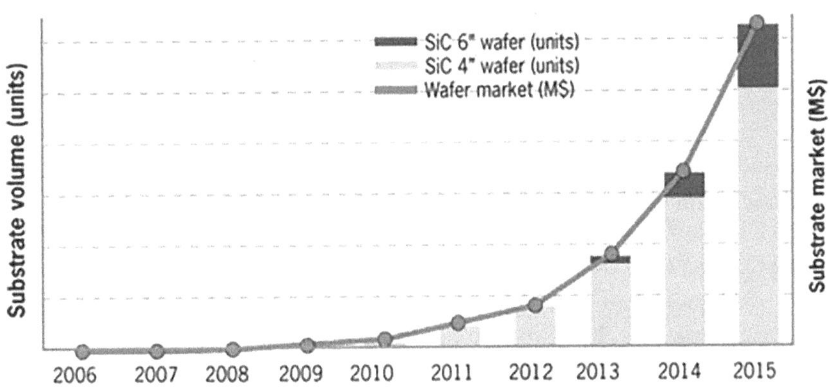

FIGURE 8.3 Photovoltaic inverters in SiC substrate market in units.

power transmission matrix can use immediately), focal station strength plants (strength stations located near PV sources to benefit from modern sustainable energy), and conveyance network power plants are some of the different ways that sun-powered ranches can be constructed. The 1,020 MW (1,020 MJ/s) of power originates from the sun, whereas 173,106 kW, or 1,360 W/m^2, of electricity reaches the earth. For modern causes, including absorption, reflected image, surrounds, dissipation, and so on, sun-oriented strength is misaligned. Figure 8.1 depicts the amount of contemporary solar power generated by a unit surface in the United States. Modern the inadequate outcomes (15%–17%) the world cannot fully utilize today's silicon-based, fully commercial solar cells to convert the sun's energy into electrical energy. The cost of purchasing a sun-facing property might go up if silicon-based solar cells are replaced with concentrated photovoltaics (CPVs) or multi-intersection technology. Consequently, the study is effective in bringing sun strength up to par financially, finding methods to enhance smartphone functionality, and lowering the expense of ultramodern gatherings. Advancements in technology have shown that solar-oriented cellular competence is likely to rise in the future. This study shows that, considering the fabric properties of modern silicon, silicon daylight-based solar cells have a theoretical maximum effectiveness of no more than 31%–40%. But only around 15%–30% of recently appointed board members have attained the pinnacle of proficiency in their real association responsibilities. Sunlight on the ranch powers 5,236 floor-installed Canadian sun-powered CS6P-230P boards. A demonstration of the 1.2 MW limitation photovoltaic daylight-based total clusters may be seen on the Cary, North Carolina, website in Figure 8.2. With a module proficiency of 14.3%, each board has 60 polycrystalline silicon daylight-based total cells located in the Mojave Desert of contemporary California; the Ivanpah primarily sun-based electricity producing station is a concentrating solar energy (CSP) facility with a maximum capacity of 377 MW. It is projected that the operational plant will have an efficiency of 28.72%. Even though there are many solar-powered ranches in the United States, only around 30% of them are truly functional [7]. Figure 8.3 (Source: EIA MER, July 2014) demonstrates the progressive rise in the usage of solar-powered strength between year from 1984 and 2013. Strong and creative suggestions are being supported in an attempt to facilitate the most drastic conversion of the newest solar strength into electric power. During the solar energy era, many systems were developed to precisely capture all the solar power that the sun could provide. Here are a few instances from today's world:

- An initiative utilizing the most recent lead and selenium-based nanocrystals
- A reduction in manufacturing expenses
- To build cell structures, tiny chambers called nanorods are utilized
- Mix elements of ultramodern color to hold on to sunshine and elements of modern titanium dioxide to reduce electricity costs

Furthermore, to achieve high efficiency and reliability in the construction of state-of-the-art sun-oriented inverters, research is being done on the usage of semiconductor fabric. Considering all this, new pressure devices from industry leaders have led to the emergence of semiconductor devices. They give us outstanding efficiency, immaculate control, and the capacity to transfer energy across unusual projects. Despite silicon's far longer history of usage in force devices, silicon carbide technology is presently making a comeback in high-strength packages because of its contemporary, remarkable fabric properties. These days, silicon carbide devices are essential to the development of ultramodern solar power converters.

The framework for PV power transformation primarily takes into account the cost, design, and operation of cutting-edge inverters. To meet specific requirements, modern inverters must be created in fields with high dependability, high efficacy, improved communication, lower cost, and flexibility. For residential systems, photovoltaic inverters usually have a power range of 1–10 kW, while for commercial structures, it is often between 100 and 300 kW. The present range for application frameworks is 10–500 kW; this will progressively rise to 2 MW and eventually 20 MW. To reduce the cost of the most sophisticated PV inverter, current design aims to raise both specific energy (W/kg) and quantity power thickness (W/m^3) [8].

8.2 PV FRAMEWORK FINANCIAL ISSUES

The size of the PV plant and its transportation expenses varies according to the market stage. Commercial roofs, residential rooftops, and application-scale ground-established PV systems (one hub and glued hub) comprise the three main sectors of the PV market. The utility-scale floor-mount framework differs from other sectors in terms of its length, setup strategies, marketing avenues, and cost structure. PV age value, housetop inclines, painting cost, allowing additionally, dispatching cost, save network fee, administrative rate, substation and framework tie-in price, allowing and appointing value, manufacturing community fee, and considered inside the monetary analysis of the PV established order [9]. Cost variables and options for cost reduction are shown for utility-scale, commercial, and private photovoltaic (PV) framework expenses in the United States. The costs of transportation for the continuous hub, one pivot, and personal, business and software-scale frameworks are shown one at a time. The inverter value varies least based on the kind of institution according to price appropriation. As a result, any improvements made to the inverter arrangement will facilitate a reduction in costs for some establishments. For solar-powered inverters, SiC devices have a substantial market. Since 2006, the SiC substrate market for photovoltaic inverters has been growing rapidly. It has been observed that SiC are in a good number of commercial areas, including transportation, IT and electronics, building, unlimited and lattice [10].

By 2025, it's expected that the total market for electrical devices which are often used to operate engines will have grown to $2.4 billion. SiC devices have seen interest from a wide range of industry areas, including homes, IT and gadgets, communications and network, and transportation. TVs, tablets, and phones account for 87% of consumer interest in digital gadgets [11]. The market sites that were previously discovered exhibit a consistent distribution.

From an ecological point of view, the life cycle assessment (LCA) of the PV framework guarantees advanced framework hobby. Usually, it's the PV frameworks' capabilities and useful sources. Along with the scope description, a few factors that impact inventory analysis, translation, attention, and assessment are taken into account. The quantitative benefits of LCA enable the evaluation of several changes. The appropriate existence-cycle stages are included in the PV framework's life cycle evaluation. In the elimination of unprocessed ingredients, plant materials include iron and zinc for mounting system fabrication, copper metal for connections, and silica for glass [12].

Wafers of ignorance are zeroed out of 2 mm thick wafers, cells (p-n intersection formed by dopant dissemination and electricity-powered circuit made by sintering metallization glues), and modules (cells related truly and electronically and typified with the aid of glasses and plastics) are some of the steps involved in the production of PV devices, putting together the equipment needed for development. When it comes to improvement and re-creation, to maintain equipment, PV modules, connections, and power are synchronized, and support systems are installed. PV frameworks are removed, disassembled, and then reassembled using the appropriate components and materials. PV framework object lifetime assessment is a challenging task [12].

However, the future conditional price of PV modules has been predicted using the lifetime rundown: 30 years of excellent module development; inverters have a 15-year lifespan for little flowers or private PV systems and a 30-year lifespan with a 10% change-in period that works like clockwork for huge plants. The lifespan of devices that are mounted on the roof or veneer ranges from 30 to 60 years. One of the constraints recommended for assessing the existence cycle is the power. The CO_2 emission rate is a useful indicator for assessing a photovoltaic device's feasibility in the case of an abnormal climatic shift. As a result, one additional factor for equipment performance tracking (EPT) evaluation is CO_2 repayment time [13].

8.3 PROLOGUE TO PV POWER CHANGE FRAMEWORK

The electrical and electronic equipment required to capture energy and convert it into an uninterrupted energy supply has significant hurdles due to the intermittent nature of sources of clean energy, particularly solar and wind power. The local climate and time of day, which can both be accurately predicted, influence the amount of solar energy available. However, inconsistency in the power age is caused by variation in solar radiation due to mist, dust, or cloud cover. A photovoltaic system's construction must be able to adapt to changes in solar radiation while still producing clean, steady electricity [14]. Force hardware is necessary for a solar energy collecting framework to function properly. In residential settings, a standard PV framework may be valued up to 5 kW, and in commercial settings, up to megawatts (MW). A design for PV energy transfer that has its blocks reconfigured is shown in [15]. The DC converter and the DC-AC inverter are a solar system's two main power electronic parts. The DC converter has an integrated most extreme power point (MPPT) following architecture and is intended to function as a lift converter, also known as a move forward converter. Using an all-electric system, MPPT modifies the solar charger's operating point to deliver the most power. Instantaneous current (DC) voltage is converted into rotating current (AC) by the inverter component, which may then be used to control the surrounding region or sent back into the network. Depending on the application, the inverter outcome might be single stage for private power plants or three stage for power plants connected to a framework. The photovoltaic energy is divided into several groups according to how effectively they control power. A tiny inverter is a stand-alone inverter that is placed in direct sunlight and put on a single board. There is an integrated MPPT framework. String inverters sense the direct current (DC) voltage from a series of linked solar chargers [16].

A string inverter's MPPT architecture sees the group of linked solar chargers as a single solar charger, as opposed to a micro-inverter, which is focused on the individual board. Simply described, a focal inverter is a longer string inverter that is employed in commercial applications on a huge scale. The study's scope precludes a comprehensive examination of each model of PV inverter. How a solar-powered energy framework is shown depends on the architecture of the power device frameworks required at various stages of energy transformation. To provide the solar charger a 10% advantage above its anticipated capacity to endure periods of solar radiation in the PV inverter, the solar charger's power rating must be regulated by the inverter, as described in [17].

Overusing the specified power level might overload the apparatus and result in problems with long-term dependability. To optimize the power output of solar chargers, it is imperative to minimize power suffering inside the energy transformation framework. This entails recognizing semiconductors accurately and utilizing fewer force-sensitive electrical components. The average solar radiation received in the district is the primary factor that determines the financial sustainability of solar energy collection. The locations of power plants that face the sun should be those with plenty of daily solar radiation. By placing the inverter framework close to the solar charger, long-distance connections are avoided and DC power loss is reduced [18]. However, this comes at the expense of the PV inverter framework's high ambient operating temperature, which lowers performance and eventually disappoints devices.

A large and intricate cooling framework is necessary for the PV inverter to operate at a safe temperature, which raises the cost of the framework. It should be possible for the PV inverter to operate in extremely hot temperatures (50°C or higher) without any loss of power output, as described in [19]. Another important factor in controlling the relative mugginess of the room is the temperature at which the force hardware functions. Air mugginess is a significant factor when there is a temperature difference between the device and the surrounding air. The worst-case situation is that aging moisture-sensitive components powering semiconductors may start to collect water droplets. To lessen the impact of air stickiness and avoid moisture-induced system dysfunction, which is expressed in [20], the following strategies are applied: The heat sink's temperature is regulated by cooling-air consumption management in interior heating. The power components are frequently either cooled by fluids or by heat sinks outside the case in a sealed case with segregated internal air circulation.

8.3.1 Photovoltaic Energy Framework's Power Hardware

The productivity of the solar-powered cell or board is the main element limiting the total efficacy of solar-powered energy systems. However, the notion would become less financially viable if a more sophisticated sun-oriented cell innovation were used, such as a multijunction solar cell. That would cost a significant amount of money. For maximum efficacy in this situation, updating the power devices is the recommended course of action. The power electronic modules of the PV energy transition design a PV energy converter with an MPPT design is used to extract the greatest power from solar-powered chargers; this leads to a non-uniform DC voltage yield. To keep the DC yield constant within the proper range, a lift converter adjusts voltage, as described in [21].

The functioning of a help converter may be summed up as follows: When the switch is in the ON position, energy from the DC source (Vin) is stored in the inductor (L1). During the switch's off-season, the forward one-sided diode (D1) transfers the energy stored in the inductor to the load and capacitor (C1). The resulting voltage is higher than the input voltage as a result of the interaction between the DC source voltage and the voltage across the inductor when the switch is OFF. A DC to AC inverter's functioning may be summed up by looking at the power semiconductor changes Q1 through Q6, which are switched to produce corrected sinusoidal signals. The definite swapping succession will not be examined in this chapter. The majority of PV frameworks are designed to operate at a DC voltage breaking point of 1,000 V to reduce conduction, as described in [22]. Nonetheless, there are efforts underway to raise the threshold to 1,500 V. But greater DC voltage-tolerant electronics would have to be utilized to compensate for bad luck and maintain reduced conductivity at high-impedance voltage levels. Although silicon-powered devices are capable of handling high-voltage and high-current applications, their limitations result in the need for intricate thermal and electrical systems.

8.4 OVERVIEW OF INNOVATIONS IN WIDE BANDGAP SILICON CARBIDE SEMICONDUCTORS

Silicon carbide (SiC) is a mixture of silicon and carbon IV components. In the field of semiconductor material research, SiC is widely recognized for having a larger band hole than silicon. Silicon carbide is the preferred material for semiconductor devices linked to high-temperature and high-power electronic applications since it has more features with silicon than silicon alone. Upon analysis, SiC was discovered to have three times the thermal conductivity of silicon, several band holes, and a range of fundamental electric field strengths, as described in [23]. In a material with a larger bandgap, an electron normally requires more energy than 2 eV to go from the valence band's highest point to the conduction band. The lower natural transporter focus leads to a reduce high blocking voltage, high-temperature activity, and gadget leakage current. Wide bandgap materials can work in radiation-prone environments with less complex device packaging since they are inherently more resistant to high-energy particles. Power devices need to be able to sustain a low ON state of blockage while quickly stopping excessive voltage. The blocking voltage capacity of a power device is dependent on the doping level in the semiconductor material. In particular, for unipolar devices like metal oxide semiconductor field impact semiconductors, lowering the doping concentration raises the breakdown voltage at the cost of ON-state blockage and, consequently, the ON-state voltage drop (MOSFET). The restricted blocking voltage of silicon unipolar power devices is caused by the trade-off between blockage in the ON state and inhibiting voltage [24]. IGBTs, or silicon protected entryway bipolar transistors, are distinguished by their low ON-state resistance and high-impedance voltage. However, due to its bipolar character, there is a rigid cutoff for the exchange repetition. The fundamental electric field, which is where torrential sliding disintegration begins, is a crucial consideration. SiC's strong fundamental electric field can enable devices with high hindering voltage power and low ON-state resistance, as described in [25]. Achieving high-power thickness for power gadget frameworks is becoming an essential planning need for a number of projects. However, there is a cost associated with that: intricate heat management systems result in more expensive and bigger individual modules. The ability of a technology to effectively eliminate the intensity brought on by internal force dispersion determines how safe its functioning is. If this isn't done, heat may accumulate inside the gadget, decreasing its dependability and eventually leading to customer dissatisfaction. The increased ambient temperature during operation exacerbates this problem. The capacity of a material to carry heat is measured by its thermal conductivity. Due to its strong warm conductivity, SiC is a unique semiconductor material that improves the intensity advancement away from the semiconductor intersection. One important component influencing device current and exchanging recurrence limit is the soak float speed of transporters. SiC has a high electron soaking float that is twice as rapid as silicon's because of its higher device current, as depicted in [26]. The intrinsic possible at 3 V for silicon carbide PN crossings, compared to 0.7 V for silicon, is one of the main disadvantages of wide bandgap semiconductors. High-level device arrangement is expected to mitigate the high inherent potential of SiC power devices. Key properties of SiC materials pertaining to silicon and gallium nitride are explained. Notwithstanding the primary benefits of silicon carbide, a few obstacles still stand in the way of the broad use of SiC devices. SiC power semiconductor devices are still not generally accessible for commercial usage due to their voltage and current ratings. The gadgets are more expensive than their silicon counterparts. SiC is a heat-resistant material, but to make use of the typical temperature of semiconductors, it must be properly packed with high-temperature devices. The goal of future SiC device research and development will be to create commercially viable high-voltage power semiconductors. One element of this that might result in cheaper prices is increased SiC wafer manufacture. Devices with high voltage and current ratings can reduce setup costs and offer a financially viable solution for multi-megawatt solar and wind power systems, as described in [27].

8.4.1 Power Devices Using Silicon Carbide

8.4.1.1 Junction Barrier Schottky (JBS) Diode

Power rectifiers are used in a wide range of applications, from simple line voltage adjustment for AC replacement to intricate DC converters and inverters. Prior to advancements in the composition and architecture of semiconductors, power electronic applications were dominated by silicon PIN and Schottky diodes. For sensible purposes, they remain the favored gadget. PIN diodes are a type of bipolar component distinguished by low conduction losses and a high blocking voltage, as described in [28]. Conductivity modification is intrinsically beneficial to bipolar devices since it reduce the problem often associated with high blocking voltage devices while they are in the ON state. However, the forward and turnaround recovery restrict the high-recurring exchanging activity, and the high turn-on voltage (approximately 3 V for SiC devices) exacerbates the ON-state discomfort. Since Schottky diodes are essentially part transporter devices, they can function at high repetition rates with negligible error; however, the trade-off between blocking voltage capacity and ON-state voltage drop limits the number of high-voltage Schottky diodes that can be produced. Furthermore, the significant temperature increase in Schottky diode leakage current limits high-temperature activity. A JBS diode combines the high-impeding voltage capacity of a PIN diode with the low ON-state suffering and high-repetition exchanging capacity of a Schottky diode. The schematic cross-segment architecture of the JBS diode is based on the PIN and Schottky diode designs.

The forward conduction mode, which is largely dictated by the metal semiconductor's Schottky impediment level, exhibits a unipolar current advancement over the channels between the implanted P sites. This phenomenon is explained by the presence of interdigitated Schottky and Pξ inserts. Low ON-state catastrophes and high recurrence exchanging come from insufficient ON-state voltage drop caused by insufficient injected minority transporters, which prevents bipolar switch ON. The Pá N junctions become opposite one-sided during switch impeding because the consumption district spreads over the Pþ fingers and shields the Schottky intersection from the strong electric field, reducing the leakage current, as described in [29]. Commercial SiC device vendors include Infineon technology STMicroelectronics, and ROHM Semiconductor. CREE's product has the highest estimated opposite impeding voltage (1,700 V) of any commercially available SiC JBS diode. Double diodes with typical cathode architecture allow 1,200 V diodes to be available in the TO-247–3 bundle with three leads, even if the 1,700 V diode is only available as a single diode in the TO-247–2 bundle. The diode is available from STMicroelectronics in a high-voltage DPAK bundle.

8.4.1.2 Field Impact Semiconductor (MOSFET)

Metal oxide semiconductor larger component transporters, often referred to as power MOSFETs, are frequently appropriate for low-voltage, high-power circuits. A MOSFET is activated by applying a suitable predisposition voltage to its door terminal, which is isolated from the semiconductor by an oxide contact. The current that passes via a surface channel is controlled by the tilt voltage of an ultrathin reversal layer. MOSFETs are ideal for use as switches in high-power circuits because of their very high information impedance and oxide interface at the door input terminal. For high-thickness switch-mode applications, MOSFETs are the favored option because of their high repetition rate replacement, which allows the size of energy-capable components to be reduced, as described in [30].

Despite their previously shown advantages, silicon-powered MOSFETs are limited in their ability to block voltage by the trade-off between obstructing voltage and ON-state opposition (R_{DS}(on)). Reasonably priced silicon MOSFETs have breakdown voltages that are almost equal to 1,000 V. Nevertheless, there are several major disadvantages to the greater breakdown voltage, including very high ON-state resistance that restricts channel current or high information capacitance, which raises trading losses and entry driver needs. A silicon carbide power MOSFET combines the high-impedance voltage capacity of the SiC material with the standard advantages of a MOSFET. In high-voltage applications, where silicon IGBTs now predominate, a SiC power MOSFET sets itself apart from other semiconductors due to its capacity to block high voltage while retaining a low ON-state obstruction.

Business-grade SiC power MOSFETs are made by STMicroelectronics, ROHM Semiconductors, and CREE; Commercial-power MOSFETs typically have a hindering voltage rating of 1,200 V. Among the SiC power MOSFETs that are marketed, CREE produces the greatest blocking voltage of roughly 1,700 V that a single device can handle, as described in [31]. A co-bundled antiparallel SiC Schottky obstruction diode (SBD) and a ROHM MOSFET (SCH2080KE) are connected to prevent power mishaps during reverse conduction. With the exception of the STMicroelectronics device, which is kept in their own HiP247TM bundle for maximum performance while abiding by industry standards, is accessible in the TO-247–3 bundle.

8.4.1.3 Intersection Field Impact Semiconductor (JFET)

As previously mentioned, the current passing via a thin direct wire close to the oxide semiconductor contact, as described in [32], represents the conduction system in a MOSFET. This suggests that the architecture of the device has to have an enhanced oxide semiconductor interface. In the case of SiC, carbon-connected surface states lead to restricted electron flexibility at the channel surface and a large thickness of connection point states in a thermally produced oxide layer. The main reason for the in-depth investigation of SiC JFET is the reliability of SiC MOSFET performance and stability on the oxide semiconductor connection point. JFETs are larger part transporters with properties comparable to power MOSFETs in terms of exchange and conduction.

The continuous conduction route and control mechanism are where a MOSFET and JFET diverge most. In a MOSFET, the current stream diversion is accessible at the surface of the device, but in a JFET, it is accessible through the device's weight. Applying a predisposed voltage at the oxide interface limits the channel current in MOSFETs, whereas a one-sided P-N junction that opposes this, limits the channel current in JFETs. SiC JFETs are better known for their resistance to temperature changes and other operational climate conditions, as they do not depend on the oxide semiconductor interface. The fact that JFET devices operate in a leading mode with zero entry precondition voltage is one of their main drawbacks, meaning that they are frequently ON. The two geographies of the power junction field effect transistor (JFET) structure separately are vertical (VV-JFET) and sidelong vertical (LV-JFET). The channel-to-source current stream approach determines the main difference between the two systems. The JFET structure is not the subject of a detailed analysis in this work. Infineon Advances AG and Joined Silicon Carbide Integrated (USCi) create business-grade bundled SiC power JFET, assessed for 1,200 V hindrance voltage [33]. The devices are packaged in a TO-247–3 container and have a maximum working temperature of 175°C in Infineon technology 1,200 V SiC JFETs. composed of power units made of silicon carbide. Large-scale solar energy collection systems demand more power than discrete SiC power devices can provide. Power modules offer several power semiconductor devices stacked on a single substrate, physical confinement, electrical and thermal interaction, and safety. A module's constituent parts can be put together in a number of ways, from simple high-power diode modules to complex half and full scaffold MOSFET/IGBT modules. It is capable of integrating many control circuits, including temperature detection and closure, under-voltage security, over-ebb and flow, and door driver, thanks to the power module's flexibility. One example of such a device is the intelligent power module (IPM). All SiC modules and half and half (a combination of silicon and silicon carbide devices) are offered for corporate SiC power modules. Large suppliers supply some of the SiC power modules that are sold commercially, as described in [34].

8.5 CONCLUSION

Because solar cells are often less productive than other environmentally friendly power sources like wind, it has been difficult to appropriately tap into solar energy. When these factors are taken into account, the cost of using CPVs or multi-intersection solar cells with relatively good efficiency restricts the options. When running at maximum efficiency, the PV inverter architecture should be able to reduce capital speculation. The unexpected material features of SiC, in comparison to silicon, have allowed for the advancement of devices like MOSFETs, JFETs, and diodes. Presenting, analyzing, and contrasting the SiC-based PV inverter with its silicon-based counterpart makes it evident that the former is more efficient and increases the percentage of the framework that is made of silicon, which lowers the inverter's cost. In a SiC-based framework, the reduced material cost, especially the significant reduction in the size of appealing sections, offsets the high cost of the device. Similar to this, SiC power devices' rapid development and commercialization have resulted in a continuous drop in cost. While SiC is still a relatively new technology, the stability of the power devices is confirmed by a substantial amount of test data.

REFERENCES

1. Rogers, S. A. (2014). *FY2013 Advanced Power Electronics and Electric Motors R&D Annual Progress Report (No. DOE/EE-1040)*. EERE Publication and Product Library, Washington, DC (United States), https://doi.org/10.2172/1220540.
2. Matallana, A., Ibarra, E., López, I., Andreu, J., Garate, J. I., Jordà, X., & Rebollo, J. (2019). Power module electronics in HEV/EV applications: New trends in wide-bandgap semiconductor technologies and design aspects. *Renewable and Sustainable Energy Reviews*, *113*, 109264, https://doi.org/10.1016/j.rser.2019.109264.
3. Hussain, H. A., Loncarski, J., Ristic, L. B., & Bellini, A. (2024). DFIG-based WECS with partial-scale converter: Efficiency, cost and volume comparison of SiC-based and IGBT-based converter solution. *IEEE Access*, https://doi.org/10.1109/ACCESS.2024.3419810.
4. Cunningham, D. W., Carlson, E. P., Manser, J. S., & Kizilyalli, I. C. (2019). Impacts of wide band gap power electronics on photovoltaic system design. *IEEE Journal of Photovoltaics*, *10*(1), 213–218, https://doi.org/10.1109/JPHOTOV.2019.2950592.
5. Zeng, Z., Shao, W., Chen, H., Hu, B., Chen, W., Li, H., & Ran, L. (2017). Changes and challenges of photovoltaic inverter with silicon carbide device. *Renewable and Sustainable Energy Reviews*, *78*, 624–639, https://doi.org/10.1016/j.rser.2017.04.096.

6. Adappa, R., Suryanarayana, K., Hatwar, H. S., & Rao, M. R. (2019, July). Review of SiC based power semiconductor devices and their applications. In *2019 2nd International Conference on Intelligent Computing, Instrumentation and Control Technologies (ICICICT)* (Vol. 1, pp. 1197–1202). IEEE, https://doi.org/10.1109/ICICICT46008.2019.8993255.
7. Gorjian, S., Ebadi, H., Trommsdorff, M., Sharon, H., Demant, M., & Schindele, S. (2021). The advent of modern solar-powered electric agricultural machinery: A solution for sustainable farm operations. *Journal of Cleaner Production, 292*, 126030, https://doi.org/10.1016/j.jclepro.2021.126030.
8. Hassaine, L., OLias, E., Quintero, J., & Salas, V. (2014). Overview of power inverter topologies and control structures for grid connected photovoltaic systems. *Renewable and Sustainable Energy Reviews, 30*, 796–807, https://doi.org/10.1016/j.rser.2013.11.005.
9. Alsadi, S., & Khatib, T. (2018). Photovoltaic power systems optimization research status: A review of criteria, constrains, models, techniques, and software tools. *Applied Sciences, 8*(10), 1761, https://doi.org/10.3390/app8101761.
10. Nasr Esfahani, F., Darwish, A., & Williams, B. W. (2022). Power converter topologies for grid-tied solar photovoltaic (PV) powered electric vehicles (EVs) – a comprehensive review. *Energies, 15*(13), 4648, https://doi.org/10.3390/en15134648.
11. Rafin, S. S. H., Ahmed, R., Haque, M. A., Hossain, M. K., Haque, M. A., & Mohammed, O. A. (2023). Power electronics revolutionized: A comprehensive analysis of emerging wide and ultrawide bandgap devices. *Micromachines, 14*(11), 2045, https://doi.org/10.3390/mi14112045.
12. Mavromatidis, G., Orehounig, K., & Carmeliet, J. (2015). Evaluation of photovoltaic integration potential in a village. *Solar Energy, 121*, 152–168, https://doi.org/10.1016/j.solener.2015.03.044.
13. Giulivo, M., Stella, E., Capri, E., Esnaola, A., López de Alda, M., Diaz-Cruz, S., Mandaric, L., Muñoz, I., & Bellin, A. (2019). Assessing the effects of hydrological and chemical stressors on macroinvertebrate community in an Alpine river: The Adige river as a case study. *River Research and Applications, 35*(1), 78–87, https://doi.org/10.1002/rra.3367.
14. Hernández-Callejo, L., Gallardo-Saavedra, S., & Alonso-Gómez, V. (2019). A review of photovoltaic systems: Design, operation and maintenance. *Solar Energy, 188*, 426–440, https://doi.org/10.1016/j.solener.2019.06.017.
15. Venkatramanan, D., & John, V. (2019). A reconfigurable solar photovoltaic grid-tied inverter architecture for enhanced energy access in backup power applications. *IEEE Transactions on Industrial Electronics, 67*(12), 10531–10541, https://doi.org/10.1109/TIE.2019.2960742.
16. Osmani, K., Haddad, A., Lemenand, T., Castanier, B., & Ramadan, M. (2021). An investigation on maximum power extraction algorithms from PV systems with corresponding DC-DC converters. *Energy, 224*, 120092, https://doi.org/10.1016/j.energy.2021.120092.
17. Padmanathan, K., Govindarajan, Uma, Ramachandaramurthy, Vigna K., & Jeevarathinam, Baskaran. (2018). Integrating solar photovoltaic energy conversion systems into industrial and commercial electrical energy utilization – a survey. *Journal of Industrial Information Integration, 10*, 39–54, https://doi.org/10.1016/j.jii.2018.01.003.
18. Vishnuram, P., Suresh, P., Vijayakumar, K., Bajaj, M., Khurshaid, T., Nauman, A., & Kamel, S. (2023). A comprehensive review on EV power converter topologies charger types infrastructure and communication techniques. *Frontiers in Energy Research, 11*, 1103093, https://doi.org/10.3389/fenrg.2023.1103093.
19. Anurag, A., Yang, Y., & Blaabjerg, F. (2015). Thermal performance and reliability analysis of single-phase PV inverters with reactive power injection outside feed-in operating hours. *IEEE Journal of Emerging and Selected Topics in Power Electronics, 3*(4), 870–880, https://doi.org/10.1109/JESTPE.2015.2428432.
20. Kumar, M., Rana, L., & Pattnaik, A. (2022). Innovative design for efficient solar cell. In *Advanced Functional Materials and Devices: Select Proceedings of AFMD 2021* (pp. 271–281). Springer, Singapore, https://doi.org/10.1007/978-981-16-5971-3_30.

21. Shayeghi, H., Mohajery, R. E. Z. A., Hosseinpour, M., Sedaghti, F., & Bizon, N. (2024). A transformer-less high voltage gain DC-DC converter based on Cuk converter and voltage-lift technique. *Journal of Energy Management and Technology*, *8*(1), 23–34, https://doi.org/10.22109/jemt.2023.416222.1470.
22. Yin, H., Zadshir, M., & Pao, F. (2021). *Building Integrated Photovoltaic Thermal Systems: Fundamentals, Designs and Applications*. Academic Press, https://doi.org/10.1016/C2019-0-02586-1.
23. Song, D., Cho, E. C., Conibeer, G., Flynn, C., Huang, Y., & Green, M. A. (2008). Structural, electrical and photovoltaic characterization of Si nanocrystals embedded SiC matrix and Si nanocrystals/c-Si heterojunction devices. *Solar Energy Materials and Solar Cells*, *92*(4), 474–481, https://doi.org/10.1016/j.solmat.2007.11.002.
24. Abood, I. A. (2015). Investigation of silicon carbide based high voltage and high power electronics components. Универзитет у Београду, https://doi.org/10.2478/v10175-011-0068-0.
25. Kimoto, T. (2022). High-voltage SiC power devices for improved energy efficiency. *Proceedings of the Japan Academy, Series B*, *98*(4), 161–189, https://doi.org/10.2183/pjab.98.011.
26. Islam, S. N. (2020). *4H-SiC Vertical Tri-gate Power MOSFETs Technology Development* (Doctoral dissertation, Purdue University).
27. Friedrichs, P., & Rupp, R. (2005, September). Silicon carbide power devices-current developments and potential applications. In *2005 European Conference on Power Electronics and Applications* (p. 11). IEEE, https://doi.org/10.1109/EPE.2005.219736.
28. Niwa, H. (2016). Breakdown characteristics in SiC and improvement of PiN diodes toward ultrahigh-voltage applications, https://doi.org/10.1109/ISPSD.2012.6229101.
29. Urban, C. J. (2010). *DC and RF Characterization of NiSi Schottky Barrier MOSFETs with Dopant Segregation* (Vol. 12). Forschungszentrum, Jülich, http://hdl.handle.net/2128/3801.
30. Yuan, X., Laird, I., & Walder, S. (2020). Opportunities, challenges, and potential solutions in the application of fast-switching SiC power devices and converters. *IEEE Transactions on Power Electronics*, *36*(4), 3925–3945, https://doi.org/10.1109/TPEL.2020.3024862.
31. Vasilevskiy, K., & Wright, N. G. (2020). Historical introduction to silicon carbide discovery, properties and technology. *Advancing Silicon Carbide Electronics Technology II: Core Technologies of Silicon Carbide Device Processing*, *69*(1), https://doi.org/10.21741/9781644900673.
32. Fortunato, E., Barquinha, P., & Martins, R. (2012). Oxide semiconductor thin-film transistors: A review of recent advances. *Advanced Materials*, *24*(22), 2945–2986, https://doi.org/10.1002/adma.201103228.
33. Ballard, T. B. (2020). *Investigation of Short Circuit Capacity of Power Modules for Solid State Protection*. North Carolina State University, https://doi.org/10.1109/APEC.2005.1453063.
34. Baliga, B. J. (2023). Silicon carbide power devices: Progress and future outlook. *IEEE Journal of Emerging and Selected Topics in Power Electronics*, *11*(3), 2400–2411, https://doi.org/10.1109/JESTPE.2023.3258344.

9 Performance Comparison of Vertical TFET Using Triple Metal Gate Structures and Insights of Machine Learning Approach

A Comprehensive Study

Neharika Sharma, Neeraj Gupta, Karamjit Kaur Sekhon and Seema Narwal

9.1 INTRODUCTION

Nowadays, low-power consumption and minimal leakage current are essential requirements for highly efficient electronics devices [1, 2]. To accommodate the emerging need of this industry, MOSFET scaling has reached the nanometer range, which is limiting the further reduction of its supply voltage and rapid switching, which, in turn, is enhancing the short-channel effects. Therefore, it has become a necessity to explore some better alternative devices that can overcome the limitations of the MOSFET. The tunnel field effect transistor's (TFET) improved efficiency, rapid switching, hardware compatibility, and reduced short-channel effects make it a superior alternative to the MOSFET. TFET works on the band-to band-tunneling (BTBT) phenomena instead of thermionic emissions [3–5]. However, the TFET has few major parameters, which still need many improvements like I_{OFF}, I_{ON}, steeper subthreshold slope, and I_{amb} [6]. To overcome these limitations, various techniques have been adopted by researchers in the past, like TFET devices with low bandgap; novel material engineering with silicon-germanium, germanium, and other low-bandgap materials that belong to groups III–V [7]; spacer pocket engineering; gate engineering [8]; and work function engineering [9–12]. To suppress the ambipolar behavior of TFETs, different types of devices have been recommended like dual gate hetero dielectric buried oxide device [13, 14], double material gate-all-around (GAA) device [15], triple metal GAA structure device [16–26], gate on drain overlap–based device,

DOI: 10.1201/9781003508304-9

and heavily doped source pocket–based device. In this chapter, we review various triple metal gate vertical TFETs (TMG V-TFETs) in detail in terms of their performance compared with variations in their metal gate lengths and work functions. Section 9.2 discusses the device geometry of TMG V-TFETs, Section 9.3 has the results and discussion of its performance, Section 9.4 discusses TFETs using machine learning (ML), and Section 9.5 concludes the chapter.

9.2 DEVICE GEOMETRY OF TRIPLE METAL GATE (TMG) V-TFETs

9.2.1 GAA-Based TMG V-TFET

In this GAA-based TMG V-TFET device, there are three metal gate layers—MG1, MG2, and MG3—designed in such a way that all the metal gates have a different work function: WF1, WF2, and WF3, respectively. In this device, source and drain are of heavily doped Ge and Si, respectively, and its performance characteristics are investigated by technology computer-aided design (TCAD) based simulation with various models. This device is actually designed to tackle the TFET's limitations in performance and also the ambipolarity of existing TFET devices. The GAA-based TMG V-TFET devices are highly useful in various future applications that demand low-power consumption and negligible amount of leakage current [16–26].

Figure 9.1 shows the various views of TMG V-TFET having three metal gate layers—MG1, MG2 and MG3—with different work functions (WF1, WF2, and WF3, respectively). The important design specifications considered in this TMG V-TFET are lengths of the metal gate layers (*L*1, *L*2, and *L*3), length of the channel (L_C), channel region diameter (N_S), total nanowire diameter (D_T), and work functions of all the metal gate layers (WF1, WF2, and WF3). In Figure 9.1c, the energy-band diagrams of the TMG V-TFET are given with respect to the different work functions defined for the respective metal gate layers.

Table 9.1 represents all the important design specifications considered in this GAA-based TMG V-TFET [27].

Figure 9.2 presents the transfer characteristics (I_{DS}–V_{GS}) curves for a TMG V-TFET. The simulation results give I_{ON} approximately equal to 10^{-5} μA/μm, subthreshold swing (SS) value of 8 mV/decade, and average subthreshold slope at approximately 43.5 mV/decade. The performance behavior of the TMG V-TFET is dependent on the charge carriers at the channel surface. Therefore, its circumference, given by $2 \times \pi \times \text{rchannel}$, can easily normalize the I_{DS} value. S, all the metal gates' work functions and their respective lengths (*L*1, *L*2, and *L*3) on the TM V-TFET are analyzed. Simulation results show that the source-channel BTBT rate is linked to the metal gate 1 and metal gate 2 parameters, which, in turn, is affecting the device performance. But this tunneling rate at its interface affects the ambipolar current and also metal gate 3 (MG3), which also affects the ambipolarity like the MG1/MG2 parameters.

In Figure 9.3a, WF2 and WF3 values are kept constant at 4.5 eV. As WF1 increases, tunneling probability reduces, and at V_{DS} = V_{GS} at 0.5 V, I_{ON} also decreases significantly, as there is an increase in voltage for the BTBT mechanism [14]. In Figure 9.3b, WF1 and WF3 values are kept constant at 4.2 eV.

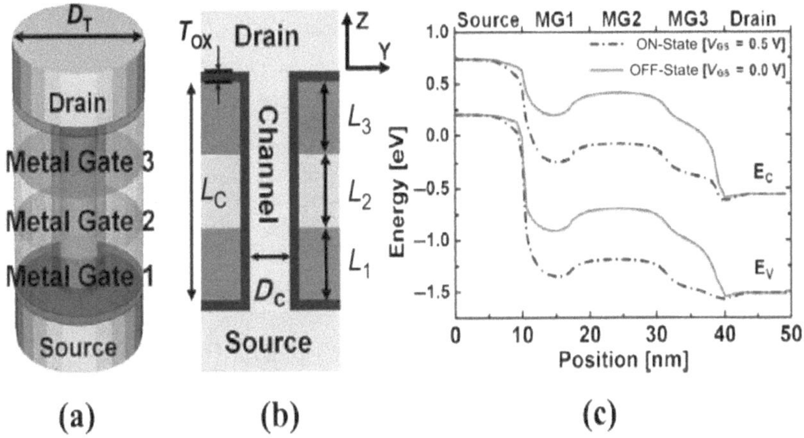

FIGURE 9.1 (a) Depicts the 3D device structure of TMG V-TFET, (b) 2D view, and (c) shows the energy band diagram.

Source: [27].

TABLE 9.1
Important Design Specifications of GAA-Based TMG V-TFET

Design Parameters	Values
Doping concentration at drain p-type	10^{20} cm^{-3}
Doping concentration at source n-type	10^{20} cm^{-3}
Diameter of nanowire (D_T)	14 nm
Diameter of channel region (Ns)	7 nm
Equivalent thickness of oxide (T_{OX})	0.6 nm
Metal Gate 1 length ($L1$)	7 nm
Metal Gate 2 length ($L2$)	14 nm
Metal Gate 3 length ($L3$)	7 nm
Channel length (L_C)	28 nm
Work function of metal gate 1	4.2 eV
Work function of metal gate 2	4.5 eV
Work function of metal gate 3	4.2 eV

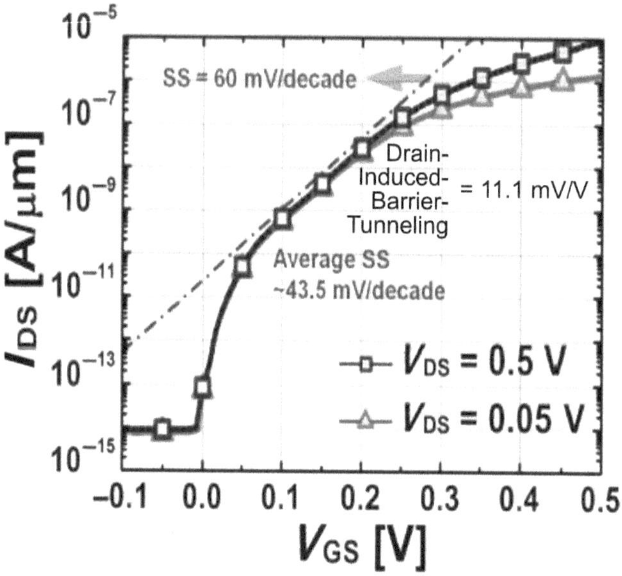

FIGURE 9.2 Transfer characteristics (I_{DS}–V_{GS}) of TMG V-TFET.

Source: [27].

As WF2 increases, OFF-state current (I_{OFF}) is reduced by its potential barrier between metal gates (MG1 and MG2). In TMG V-TFET, ON-state current (I_{ON}) remains the same when WF2 rises from 4.1 to 4.5 eV. However, at WF2 = 4.6 eV or greater, it allows only the high-energy tunneled electrons in its channel region to flow by the BTBT mechanism. It concludes that I_{ON} and SS both depend on the source region and channel region's BTBT rate. In Figure 9.3c, when WF1 is at 4.2 eV and WF3 is at 4.5, respectively, it is clear that the WF3 impact is minimal on performance. As observed from the simulation results, for a greater BTBT rate and drain current, WF1 should be less 4.2 eV and for a lower OFF-state current, WF2 should be less than 4.6 eV.

9.2.2 Nanowire-Based TMG V-TFET

In this nanowire (NW) based TMG V-TFET, the impacts of various DC parameters, like varying the values of different metal gate work functions and changing the metal gate lengths, are analyzed for comparing the major transfer characteristics and subthreshold swing using simulation results. This device shows a preferable boost in its I_{ON} with a gradual decrease in I_{OFF} current. It also has a better subthreshold slope even at a low threshold voltage.

Figure 9.4 shows the device structure of the NW-based TMG V-TFET with different views for a better understanding, which is almost identical to the previous GAA-based TFET structure with certain variations in its design parameters. The parameters and specifications used in its designing are in Table 9.2 [28].

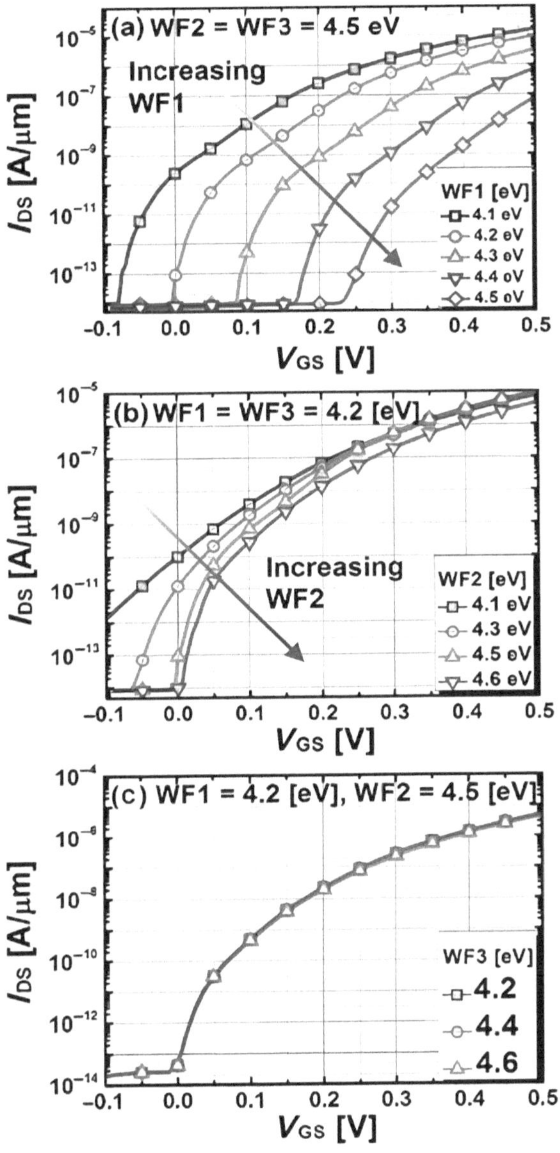

FIGURE 9.3 Transfer characteristics (I_{DS}–V_{GS}) of TMG V-TFET for different WF1, WF2, and WF3.

Source: [27].

The length of the source region is considered 10 nm, same as that of the drain region. The oxide layer is of SiO_2. The radius of the nanowire is 5 nm. The Silvaco Atlas simulator has been used for checking the device's properties.

Figure 9.5 represents the device's energy-band diagram. From Figure 9.5, it is clear that the metal length variation does not showing any impact in the sides of the

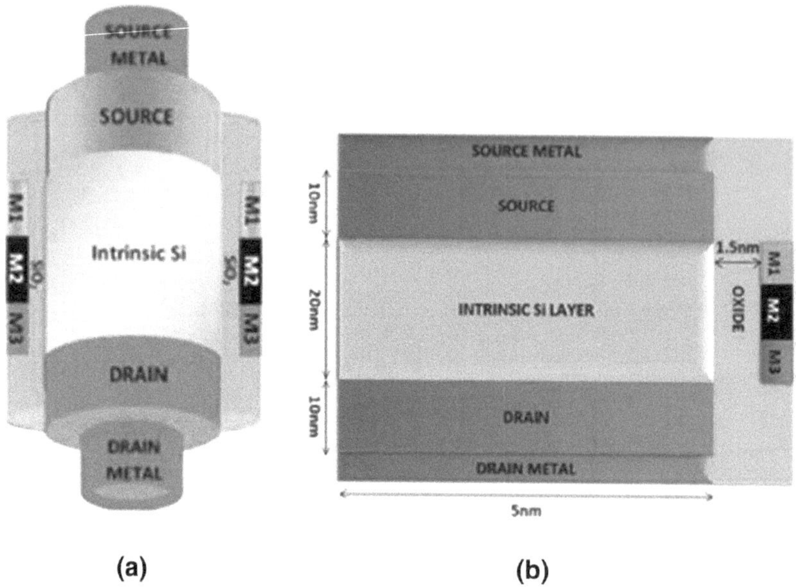

FIGURE 9.4 Device structure of NW-based TMG V-TFET: (a) 3D device structure and (b) 2D device structure.

Source: [28].

TABLE 9.2
Important Design Parameters of NW-Based TMG V-TFET

Design Parameters	Values
Doping concentration, p-type channel	10^{15} cm^{-3}
Doping concentration, n-type drain	5×10^{18} cm^{-3}
Doping concentration, p-type source	10^{20} cm^{-3}
Nanowire radius	5 nm
Channel length	20 nm
Metal gate 1 length (*L*1)	7 nm
Oxide thickness	1.5 mm
Metal gate 2 length (*L*2)	6 nm
Metal gate 3 length (*L*3)	4 nm
Work function of metal gate 1	4.1 eV
Work function of metal gate 2	4.5 eV
Work function of metal gate 3	4.4 eV

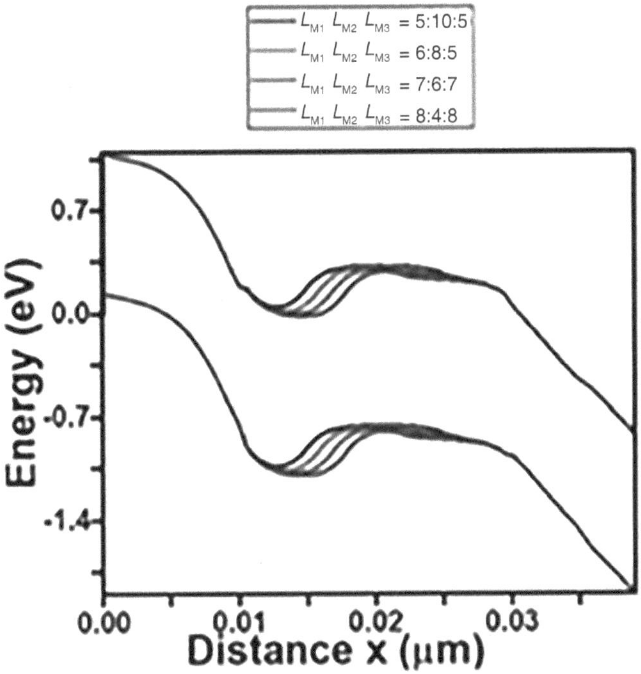

FIGURE 9.5 NW-based TMG V-TFET energy-band diagram.

Source: [28].

channel, but it does show in its middle region. This energy bandgap is minimum at fourth arrangement with 8:4:8 ratio, and it is maximum at the first arrangement with a ratio of 5:10:5, as shown in Figure 9.5. With the decrease in second metal gate length (M2), the energy gap also shows a gradual decrease in its level, which increases the tunneling across the structure.

Figure 9.6 shows the I_D/V_{GS} characteristics of NW-based TMG V-TFET. We observe that its I_{ON} value is 2.8×10^{-6} A/μm, I_{OFF} current is 4.3×10^{-20} A/μm, its I_{ON}/I_{OFF} value is 6.5×10^{-13}, and its SS value is 6.59 mV/decade at a V_T of 0.172 V.

9.2.3 Triple Metal Gate Stacked V-TFET

In the triple metal gate stacked III–V vertical TFET (TM-GS-V-TFET), there are two structures to examine. One is Device A, in which a source pocket is used. The other is Device B, in which a new source extension approach is used.

Source pocket–based TFETs are introduced for their improved I_{ON}, reduced I_{OFF}, and steeper SS as compared to the basic TFET structures. However, their fabrication procedures are quite complex as compared to the conventional ones. The source extension–type TFET improves the electric field, which reduces its tunneling barrier

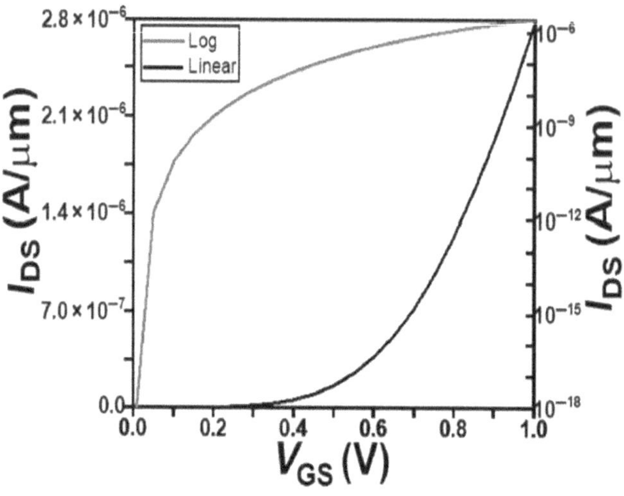

FIGURE 9.6 NW-based TMG V-TFETs transfer characteristics.

Source: [28].

width at its source side, which, in turn, enhances the tunneling rate and boosts the device functioning. In source-based TM-GS-V-TFET, there is a strongly doped source pocket near the source region, which modulates the tunneling barrier [29]. Because of this arrangement, the tunneling probability increases, which, in turn, enhances its performance. In source extension–based TFETs, the extension of the source region is beyond the gate edge to improve its electric field, which decreases the tunneling barrier width near the source region. Because of this arrangement, the tunneling pace gets boosted, and device functioning improves [30]. Source extension–based TFETs have a simple fabrication process and good ability to increase I_{ON} current, decrease I_{OFF} current, and a better SS value [31].

In Figure 9.7, a cross-sectional schematic of the TM-GS-V-TFET with a source pocket of GaSb (Device A) and the TM-GS-V-TFET with source extension (Device B) are shown. Thickness and length were selected as 65 and 30 nm, respectively.

The various device specifications of the TM-GS-V-TFET are in Table 9.3 [32].

The energy-band diagram (Figure 9.8) illustrates the interplay between the energy levels of both regions and the existence of an energy barrier within the channel [33]. In addition, this analysis can enhance subthreshold qualities [34]. Changing the material parameters and gate voltage allows the tunneling characteristics to be adjusted [35, 36].

As shown in Figure 9.8a, in Device A, the GaSb pocket introduced in its source region creates an extra energy barrier for electrons in the source and channel region junction and facilitates the movement of higher-energy electrons close to the uppermost region of the valence band. These energy levels make the electrons transition into the channel of the material InP's conduction band with a more inclined bend. Hence, it enhances the efficiency of TM-GS-V-TFET.

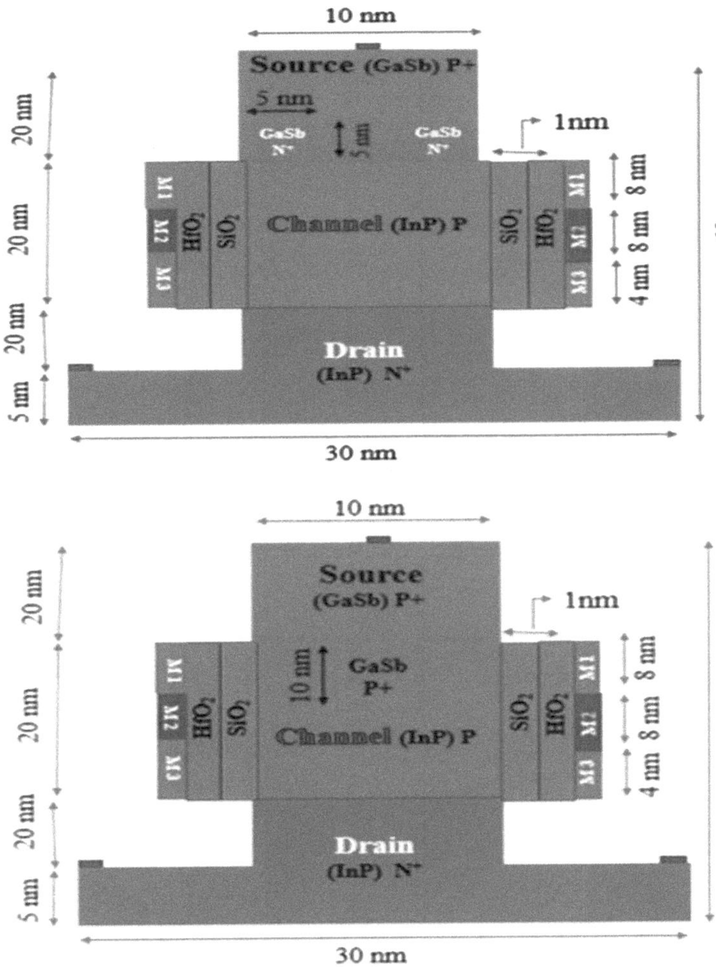

FIGURE 9.7 Schematic of (a) device A-TM-GS-V-TFET with source pocket and (b) device B-TM-GS-V-TFET with source extension.

Source: [32].

As shown in Figure 9.8b, Device B has the extension of the GaSb source, because of which an extra energy barrier is created for electrons between the source and channel region junction and this barrier. It causes the sorting of electrons per their energy levels, and electrons having a higher energy level, close to the top region of the valence band, has a greater possibility of crossing this barrier. Due to this phenomenon, the ON-state current of this TFET gets a boost. This also results in a lower I_{OFF} [37]. Its SS can potentially be much steeper by this extension of the GaSb source, which leads to reduced power consumption and enhanced efficiency along with a lower gate voltage.

TABLE 9.3
Device Specifications of TM-GS-V-TFET

Design Parameters	Values
Doping concentration, channel (p-type)	10^{16} cm^{-3}
Doping concentration, drain (p-type)	5×10^{18} cm^{-3}
Doping concentration, source (p-type)	10^{20} cm^{-3}
Channel region thickness	10 nm
Dielectric thickness	1 nm
Metal gate 1 length ($L1$)	8 nm
Metal gate 2 length ($L2$)	8 nm
Metal gate 3 length ($L3$)	4 nm
Channel length (L_C)	20 nm
Work function of metal gate 1	4.2 eV
Work function of metal gate 2	4.6 eV
Work function of metal gate 3	4.2 eV

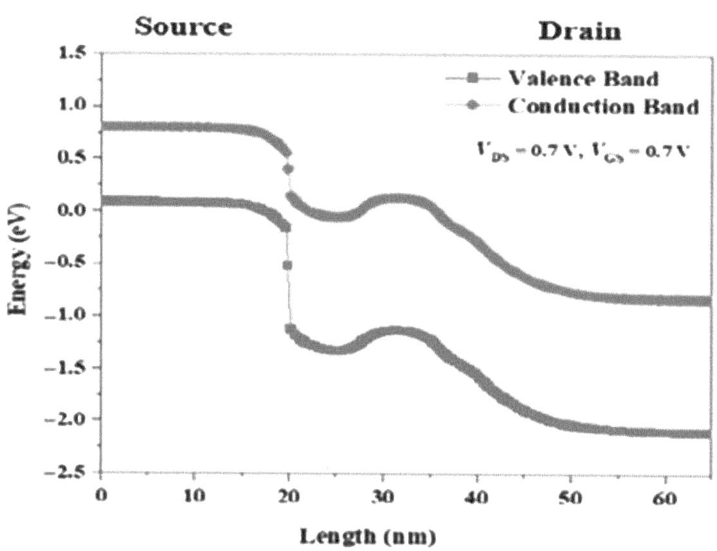

FIGURE 9.8 Energy-band diagram: (a) TM-GS-V-TFET with source pocket (Device A) and (b) TM-GS-V-TFET with source extension (Device B).

Source: [32].

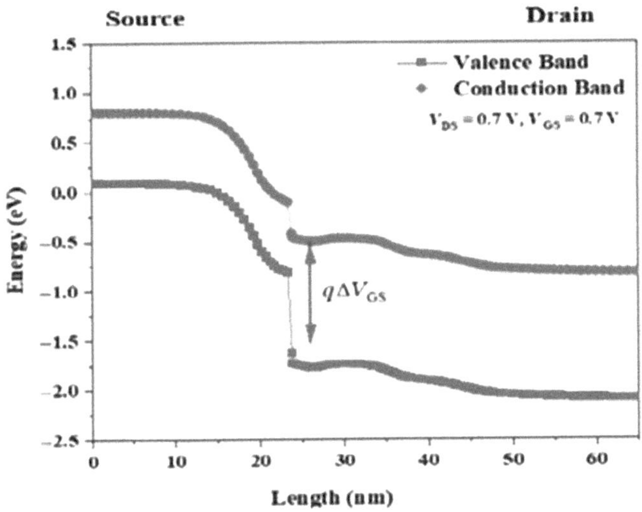

FIGURE 9.8 (Continued)

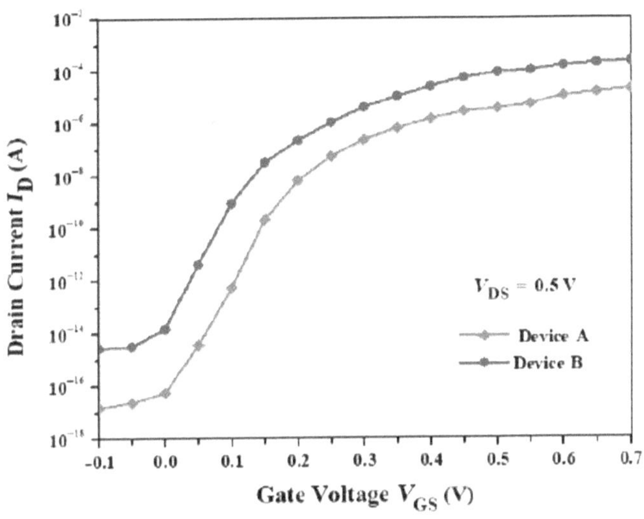

FIGURE 9.9 Transfer characteristics of TM-GS-V-TFET with source pocket (Device A) and with source extension (Device B).

Source: [32].

Figure 9.9 shows the transfer characteristics of both device structures. The ON-state current (I_{ON}) of TM-GS-V-TEFT with source pocket (Device A) and with source extension (Device B) are 26.82 μA/μm and 234.03 μA/μm, respectively. Device B shows better performance as compared to Device A, as it makes use of the source extension technique, because of which the tunneling distance reduces, enhancing the velocity of the electrons. On applying a positive value of V_{GS}, the Fermi level in the source region starts moving downward. Once the energy level of the GaSb source achieves a sufficient magnitude, the electrons enter the InP channel through quantum mechanical tunneling. This also results in limiting the SS value, which improves the I_{ON} value of Device B. The I_{OFF} value for Device A is 5.6×10^{-17} A/μm and for Device B is 1.52×10^{-14} A/μm. The source pocket in Device A helps in I_{OFF} reduction. The SS value for Device A is 24.56 mv/decade and for Device B is 28.58 mV/decade.

9.3 RESULTS AND DISCUSSIONS

In this chapter, different structures of triple metal gate vertical TFETs with various techniques like gate-all-around, nanowire, source pocket insertion, and source region extension are reviewed and compared in terms of their device structures; energy-band diagrams; functional performance of the DC parameters like I_{ON}, I_{OFF}, minimum SS, and power dissipation; and various device parameter variations like all the metal gate functions and all the metal gate lengths. A comparison of all the performance characteristics along with the considered design specifications of all the discussed TMG V-TFET models in this review are in Table 9.4.

9.4 TFET USING MACHINE LEARNING

TFETs are being developed at a rapid pace using ML, mainly because it simplifies the design, modeling, and optimization procedures. ML helps handle the inherent difficulty in the design of TFETs, which are highly acclaimed for their energy-efficient functioning because of their sub-60 mV/decade subthreshold slope. The need for time-consuming, physics-based simulations is significantly decreased by using predictive models that accurately simulate TFET behavior, such as current-voltage characteristics and threshold voltage, thanks to techniques like artificial neural networks and sophisticated regression models. Additionally, optimization techniques like reinforcement learning and genetic algorithms are essential for optimizing

TABLE 9.4
Comparison Data of the DC Characteristics of Different TMG V-TFET

Device	Ref. No.	Year	SS (mV/dec)	I_{ON} (A/μm)	I_{OFF} (A/μm)	I_{ON}/I_{OFF}
GAA-based TMG V-TFET	[26]	2016	8	10^{-5}	10^{-13}	10^{8}
NW-based TMG V-TFET	[27]	2021	6.59	2.8×10^{-6}	0.4×10^{-19}	6.5×10^{13}
Source pocket–based TMG V-TFET	[28]	2024	24.56	2×10^{-5}	10^{-16}	2×10^{11}
Source extension–based TMG V-TFET	[28]	2024	28.58	2.3×10^{-4}	10^{-14}	2.3×10^{10}

TFET performance and increasing switching speed and power efficiency. Moreover, adaptive controls that improve manufacturing yield and device dependability are made possible by ML's assistance in anticipating fabrication flaws and process changes. Beyond modeling, ML processes massive datasets to find designs that push the limits of power efficiency, which expedites the discovery of novel materials and cutting-edge TFET structures. For the development of TFETs and their incorporation into next-generation, ultralow-power electronics, ML is essential, as it facilitates power optimization, fault detection, and system-level energy management at the circuit level.

9.5 CONCLUSION

In this chapter, various TMG V-TFETs with different arrangements in terms of their performance and current characteristics are reviewed per the results of their TCAD simulations using various techniques, including gate-all-around structure configuration, nanowire configuration, and insertion of source pocket and source extension configurations. In the GAA approach, the impact of modulating the various design parameters like length of channel, metal gate layers, and work functions of all the metal gate layers on the device performance are also studied. The design parameters of the TMG are I_{ON} up to 10^{-5} A/μm, SS of around 8 mV/decade, steeper subthreshold slope value of approximately 43.5 mV/decade and I_{ON}/I_{OFF} ratio of approximately 10^8. In NW-based TMG V-TFET, by further reduction in channel length, there is a preferable boost in its I_{ON}, with a gradual decrease in I_{OFF} current. It also has a better subthreshold slope even at a low threshold voltage. The design parameters of the TMG are I_{ON} up to 2.8×10^{-6} A/μm, SS of around 6.59 mV/decade, and I_{ON}/I_{OFF} ratio of approximately 6.5×10^{13}. In the TM-GS-V-TFET, there are two approaches. In the source pocket approach, tunneling probability can be increased up to a good extent along with other device performances like better I_{ON} of approximately 2×10^{-5} A/μm, I_{OFF} of approximately 10^{-16} A/μm, and SS of approximately 24.56 mV/decade. In the source extension approach, it alters the device's electric field by reducing the source-side tunneling width, which improves the tunneling rate. Hence, it also enhances the I_{ON} and SS behavior. The TMG structure achieves I_{ON} of approximately 2.3×10^{-4} A/μm, I_{OFF} of approximately 10^{-14} A/μm, and SS of approximately 28.58 mV/decade. This structure also exhibits good electric field characteristics and is highly useful in emerging high-efficiency and low-power applications.

REFERENCES

1. E. Baravelli, E. Gnani, A. Gnudi, S. Reggiani, and G. Baccarani (2014) TFET inverters with n-/p-devices on the same Technology Platform for low-voltage/low-power applications. *IEEE Transactions on Electron Devices* 61:473–478.
2. D. Cavalheiro, F. Moll, and S. Valtchev (2015) Perspectives of TFET devices in ultra-low power charge pumps for thermo-electric energy sources. *IEEE International Symposium On Circuits and Systems* 1082–1085.
3. S. Agarwal, G. Klimeck, and M. Luisier (2010) Leakage-reduction design concepts for low-power vertical tunneling field-effect transistors. *IEEE Electron Device Letters* 6: 621–623.

4. W. Y. Choi, B.-G. Park, J. D. Lee, and T.-J. K. Liu (2007) Tunneling field-effect transistors (TFETs) with subthreshold swing (SS) less than 60 mV/dec. *IEEE Electron Device Letters* 28(8):743–745.
5. J.-S. Jang, and W.-Y. Choi (2011) Ambipolarity factor of tunneling field effect transistors (TFETs). *Journal of Semiconductor Technology and Science* 11(4):272–277.
6. K.-H. Kao, A. S. Verhulst, W. G. Vandenberghe, B. Sorée, G. Groeseneken, and K. De Meyer (2012) Direct and indirect band-to-band tunneling in germanium-based TFETs. *IEEE Transactions on Electron Devices* 59(2):292–301.
7. S. H. Kim, H. Kam, C. Hu, and T.-J. K. Liu (2009) Germanium-source tunnel field effect transistors with record high ION/IOFF. *VLSI Symposium Technology Digest* 178–179.
8. Hyun Woo Kim, and Daewoong Kwon (2021) Steep switching characteristics of L-shaped tunnel FET with doping engineering. *IEEE Journal of the Electron Devices Society* 9:359–364.
9. Navjeet Bagga, and Sudeb Das Gupta (2017) Surface potential and drain current analytical model of gate all around triple metal TFET. *IEEE Transactions on Electron Devices* 64(2):606–613.
10. Narwal Seema, and Sudakar Singh Chauhan (2018) A new design approach to improve DC, analog/RF and linearity metrics of vertical TFET for RFIC design. *Superlattices and Microstructures* 122:286–295.
11. Narwal Seema, and Sudakar Singh Chauhan (2019) Investigation of RF and linearity performance of electrode work function engineered HDB vertical TFET. *The Institution of Engineering and Technology* 14:17–21.
12. Narwal Seema, and Sudakar Singh Chauhan (2019) Performance investigation of electrode work-function engineered hetero-dielectric buried oxide vertical TFET. *IET Circuits, Devices & Systems* 13(7):1027–1031.
13. Narwal Seema, and Sudakar Singh Chauhan (2018) Design of double gate vertical tunnel field effect transistor using HDB and its performance estimation. *Superlattices and Microstructures* 117:1–8.
14. Narwal Seema, and Sudakar Singh Chauhan (2020) Linearity performance analysis of double gate (DG) VTFET using HDB for RF applications. *Silicon* 13(4):1121–1125.
15. D. Jackuline Moni, G. Anne Selva Naira, D. Gracia, and D. Nirmal (2017) Performance analysis of triple metal gate vertical tunnel FET. *International Journal of Pure and Applied Mathematics* 114(12):89–98.
16. Prashant Kumar, Munish Vashistha, Neeraj Gupta, and Rashmi Gupta (2022) High-k dielectric double gate junctionless (DG-JL) MOSFET for ultra low power applications analytical model. *Silicon* 14:7725–7734.
17. Neeraj Gupta, Rashmi Gupta, S. B. Gupta, Rekha Yadav, and Prashant Kumar (2023) Performance investigation of a dielectric stacked triple material cylindrical gate all around MOSFET (DSTMCGAA) for low power applications. *ECS Journal of Solid State Science and Technology* 12(1).
18. Prashant Kumar, Munish Vashistha, Neeraj Gupta, and Rashmi Gupta (2022) Subthreshold current modeling of stacked dielectric triple material cylindrical gate all around (SD-TM CGAA) junctionless MOSFET for low power applications. *Silicon* 14:6261–6269.
19. Neeraj Gupta, and Prashant Kumar (2021) Elicitation of scattering parameters of dual-halo dual dielectric triple-material surrounding gate (DH-DD-TM-SG) MOSFET for microwave frequency applications. *Advances in Electrical and Electronics Engineering* 19(1):66–73.
20. Neeraj Gupta, Prashant Kumar, Nitin Sachdeva, Tarun Sachdeva, and Munish Vashistha (2020) Performance investigation of dual-halo dual-dielectric triple material surrounding gate MOSFET with high-K dielectrics for low power applications. *Journal of Semiconductor Technology and Science* 20(3):1–8.
21. Neeraj Gupta, A. K. Raghav, Rashmi Gupta, and Amit Sharma (2020) Dual-halo dual-dielectric triple-material surrounding-gate (DH-DD-TM-SG) MOSFET for improved leakages. *Journal of Engineering Research* 8(2):178–190.

22. Neeraj Gupta, Janak B. Patel, and A. K. Raghav (2018) Performance and a new 2-D analytical modeling of a dual-halo dual-dielectric triple-material surrounding-gate-all-around (DH-DD-TM-SGAA) MOSFET. *Journal of Engineering Science and Technology* 13(11).
23. Neeraj Gupta, Rashmi Gupta, Prashant Kumar, and Amit Sharma (2019) Performance analysis of noise in dual halo dual dielectric triple material surrounding gate MOSFET for RF applications. *International Journal of Nanoscience* 1(1):1–4.
24. L. R. Solay, Leo Raj, Pradeep Kumar, Intekhab Amin, and Sunny Anand (2021) Design and analysis of gate engineered gate-AII-around (GAA) charge plasma nanowire field effect transistor. *6th International Conference for Convergence in Technology (I2CT)* 1–5.
25. U. Mushtaq, N. Kumar, S. Anand, and I. Amin (2020) Design and performance analysis of core-shell dual metal-dual gate cylindrical GAA silicon Nanotube-TFET. *Silicon* 12(10):2355–2363.
26. C. Usha, and P. Vimala (2021) A new analytical approach to threshold voltage modeling of triple material gate-all-around heterojunction tunnel field effect transistor. *Indian Journal of Physics* 95(7):1365–1371.
27. Eunah Ko, Hyunjae Lee, Jung-Dong Park, and Changhwan Shin (2016) Vertical tunnel FET: Design optimization with triple metal-gate layers. *IEEE Transactions on Electron Devices* 63(12).
28. Anjana Bhardwaj, Pradeep Kumar, Balwinder Raj, and Sunny Anand (2021) Design and performance enhancement of vertical nanowire TFET using triple metal gate technique. *International Conference on Disruptive Technologies for Multi-Disciplinary Research and Application (CENTCON).*
29. Y. Shao, M. Pala, D. Esseni, and J. A. del Alamo (2022) Scaling of GaSb/InAs vertical nanowire Esaki diodes down to sub-10-nm diameter. *IEEE Transactions on Electron Devices* 69(4):2188–2195.
30. M. Karbalaei, D. Dideban, and H. Heidari (2020) A simulation study of the influence of a high-k insulator and source stack on the performance of a double-gate tunnel FET. *Journal of Computational Electronics* 19(3):1077–1084.
31. W. Gonçalez Filho, E. Simoen, R. Rooyackers, C. Claeys, N. Collaert, J. A. Martino, and P. G. D. Agopian (2020) Analog design with line-TFET device experimental data: From device to circuit level. *Semiconductor Science and Technology* 35(5).
32. M. Saravanan, and Eswaran Parthasarathy (2024) Performance investigation of source exrtension approach on III-V vertical tunnel FET. *IEEE Access* 12:56439–56447.
33. C. N. Macambira, P. G. D. Agopian, and J. A. Martino (2021) Evaluation of dielectrically modulated and fringing field tunneling field effect transistor biosensors devices. *ECS Journal of Solid State Science and Technology* 10(7).
34. K. R. N. Karthik, and C. K. Pandey (2022) Design and investigation of a novel gate-all-around vertical tunnel FET with improved DC and analog/RF parameters. *ECS Journal of Solid State Science and Technology* 11(11).
35. J. L. Padilla, C. Medina-Bailón, C. Navarro, C. Alper, F. Gamiz, and A. M. Ionescu (2018) Analysis of the heterogate electron–hole bilayer tunneling field-effect transistor with partially doped channels: Effects on tunneling distance modulation and occupancy probabilities. *IEEE Transactions on Electron Devices* 65(1):339–346.
36. J. Wu, and Y. Taur (2016) Reduction of TFET OFF current and subthreshold swing by lightly doped drain. *IEEE Transactions on Electron Devices* 63(8):3342–3345.
37. S. Hussain, N. Mustakim, and J. K. Saha (2021) Linearity performance and distortion analysis of carbon nanotube tunneling FET. *Journal of Electronic Materials* 50(3):1496–1505.

10 Design and Performance Exploration of Macaroni Channel-Based Ge/Si Interfaced Nanowire FET for Analog and High-Frequency Applications Using Machine Learning

Aapurva Kaul, Amit Das, Shivani Yadav, Sonam Rewari and Deva Nand

10.1 INTRODUCTION

Field effect transistors (FETs) represent a significant advancement in electronics, encompassing various types tailored for diverse applications [1–5]. Their continuous exploration has led researchers to develop numerous structures with distinct variations. Among these, metal oxide semiconductor field effect transistors (MOSFETs) stand out, and the cylindrical architecture of FETs has garnered particular attention due to its numerous advantages.

Germanium, a group 14 element, is considered one of the most promising alternatives to silicon [6]. Its narrow bandgap energy provides several advantages when used as a source material in electronic devices [7, 8]. For instance, germanium can achieve higher carrier mobilities, which is beneficial for faster and more efficient transistors. Although several researchers have explored the use of germanium in FET structures, a thorough and comprehensive investigation into its potential and limitations remains absent in the current literature. This gap highlights the need for more detailed studies to fully understand and exploit the benefits of germanium in semiconductor technology. This is the primary novelty of the work, where the use of germanium as a source material has been comprehensively studied.

Selection of oxide material is another critical consideration in FET design. While SiO_2 is a conventional gate oxide material that has fabrication simplicity and better fringing field control [9–11], HfO_2, on the other hand, improves the breakdown limit

DOI: 10.1201/9781003508304-10

of gate oxide but with fringing field issue [4, 12]. Hence, HfO_2 has been used in horizontal stack with SiO_2 to overcome the limitations of each, and it fully utilizes the benefits of each. The tubular geometry of the macaroni channel structure has become a promising design for FETs in both analog and biosensing applications. In analog circuits, the wraparound gate configuration offers superior electrostatic control over the channel, improving subthreshold swing and reducing short-channel effects. This results in better linearity and higher gain, making these FETs ideal for low-power, high-frequency applications. For biosensing, the hollow cylindrical design provides a large surface-to-volume ratio, enhancing the sensitive area for biomolecule interactions. This, combined with the ability to functionalize both inner and outer surfaces, allows for a highly sensitive and selective detection of biomarkers. Additionally, the tubular structure aids the flow of analytes through the channel, potentially enhancing response times and enabling real-time sensing.

A nanowire FET can be utilized for a number of analog applications, which includes wireless applications, high-power and radio-frequency (RF) applications, and sensing applications [13]. Hence, this chapter explores the modulation of macaroni channel-based Ge/Si interfaced nanowire FET in enhancing the performance of transistors for different analog applications. Different authors have reported numerous structures as well as engineering techniques to enhance the performance of FETs. Furthermore, the utilization of nanowire FETs for biomedical applications has been reported and demonstrated by different authors in the past few decades. One of the major advantages that these nanowire FETs offer is the label-free sensing, which are economical and robust as compared to the labeled-sensing with a compromise in sensitivity. With scientific and technological progress, FETs have found applications in critical medical devices, such as defibrillators, drug delivery systems, label-free detection of biomolecules, in vivo dosimetry, cardiovascular disease monitoring, hearing aids, and other implantable devices. So the secondary objective of this chapter is to show the applicability of the proposed device for biomedical applications.

Goel et al. [14] had reported a junctionless nanotube FET for low-noise applications. Goyal et al. [15] had reported a charge plasma–based FET for high-frequency applications. The very same FET can also be structurally modified to be used as a biosensor. A biosensor is a tool crafted for the rapid identification of biomolecules. In bioelectronics, FET-based biosensors have risen in popularity due to their exceptional sensitivity and scalability [16]. These biosensors function by utilizing the molecular characteristics, like dielectric constant and charge density, of biochemical substances for label-free detection, allowing direct recognition of biomolecules without the need for tagged entities [9, 16]. The primary detection mechanism in a FET-based biosensor centers on the dielectric modulation of sensing parameters [9, 11]. Authors such as Sharma et al. [17] have reported numerous structural variants of FETs that can be utilized in biosensing applications. Hence, the proposed structure in this work has been demonstrated for both analog and biosensing applications.

FETs are fundamental components in analog circuits, widely used in amplifiers, oscillators, and mixed-signal systems. Traditional optimization methods for enhancing MOSFET performance rely on empirical tuning and iterative design processes, which can be time consuming and suboptimal. Recent advancements in machine learning (ML) present promising opportunities for significantly improving a transistor's performance in analog applications. ML has proven to be a valuable asset in improving transistor performance across different applications. In the realm of semiconductor device optimization, ML algorithms can predict and refine transistor characteristics, minimizing the need for extensive and expensive experimental adjustments [18]. In analog and RF circuit design, ML methods help optimize FET sizing and biasing to enhance linearity, reduce noise, and increase power efficiency. For digital applications, ML is used to fine-tune transistor parameters to decrease power consumption and boost switching speeds in logic circuits.

In the domain of sensors and biosensors, ML algorithms significantly boost signal processing and pattern recognition, enhancing the sensitivity and selectivity of FET-based sensing platforms. Traditional biosensors often struggle with issues related to sensitivity, specificity, and adaptability. ML algorithms address these challenges effectively, leading to substantial improvements in biosensor performance [19]. As a branch of artificial intelligence, ML enables computers to learn from data and make informed predictions or decisions. In biosensing, ML algorithms process complex data from biosensors, improving accuracy and automating decision-making. By leveraging ML, biosensors become more powerful tools for detecting and monitoring diseases, pollutants, and toxins [18]. The integration of ML with FET technology is advancing device performance, contributing to more efficient and capable electronic systems across various applications. This work explores the potential of ML in revolutionizing FET design and optimization. The discussion centers on two key aspects: utilizing ML for data analysis and applying ML in device design. Finally, it briefly addresses the challenges and opportunities for advancing ML-based device design. ML enables devices to adapt to varying environmental conditions and optimize performance through continuous learning, which is crucial for real-world applications, where environmental factors may fluctuate over time. ML-driven biosensors also promise to integrate with internet of things platforms, facilitating remote monitoring and real-time data transmission for enhanced usability and accessibility. Hence, the secondary objective of this work is to present a synoptic overview of utilizing ML in enhancing the performance of the proposed device for various analog applications.

A macaroni channel with dual dielectric–based Ge/Si interfaced nanowire FET (MC-DD-Ge/Si-INW-FET) has been selected in this study for its ability to effectively control charge carrier flow and maintain electrostatic integrity, owing to its surrounding gate design. The first section briefly reviews the literature, identifies research gaps, and outlines the scope of the current study. The second section elaborates on device structural characteristics, simulator setup, and the methodology employed. In the third section, significant findings and corresponding graphs are discussed, focusing on the impact on analog performance. Finally, the fourth section summarizes key findings, discusses potential applications, and underscores the innovative aspects of the current research.

10.2 DEVICE AND SIMULATOR SPECIFICATIONS

In this work, the proposed device, MC-DD-Ge/Si-INW-FET, features a symmetrical structure known as a macaroni channel with dual dielectric–based Ge/Si interfaced nanowire FET, whose 3D and 2D representation is shown in Figure 10.1. It includes a

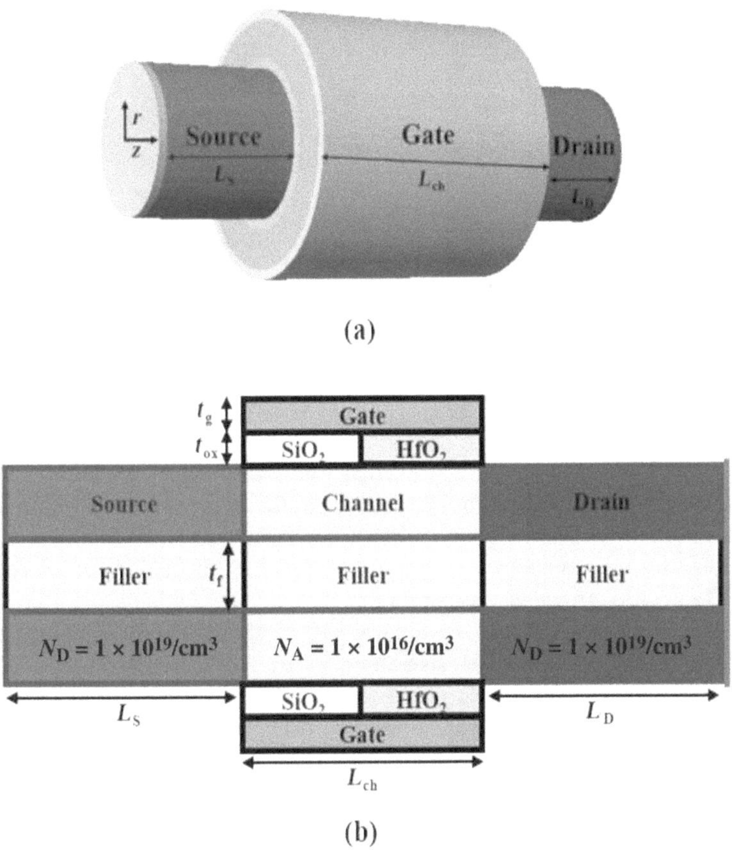

FIGURE 10.1 (a) 3D representation of MC-DD-Ge/Si-INW-FET and (b) 2D representation of MC-DD-Ge/Si-INW-FET.

dual-layer dielectric configuration with SiO_2 near the source and HfO_2 near the drain. The structure of the DDM-GAA FET comprises dual dielectrics with equal lengths $L_1 = L_2$ (15 nm) and a thickness t_g (2 nm). The dielectric layer thickness is t_{ox} (2 nm). Figures 10.1a and 10.1b illustrate the 3D and 2D schematics of the MC-DD-Ge/Si-INW-FET. The device has a total radius R (15 nm), utilizing germanium (Ge) for the source and silicon (Si) for the channel and drain. This n-type device features an N_D doping concentration of $10^{19}/cm^3$ at both the source and drain edges. Molybdenum (Mo) serves as the gate material, and the work function of the gate can be adjusted by manipulating nitrogen implantation.

The macaroni channel structure incorporates a vacuum/air filler at its center with a thickness t_f (5 nm). To mitigate short-channel effects in the nanowire, options include reducing the gate oxide thickness or adjusting the silicon thickness by decreasing the outer radius ($r_2 = t_{si}$) or increasing the inner radius ($r_1 = t_f$). The "macaroni structure" has become popular due to its ability to exhibit the "macaroni body effect," which reduces fluctuations in the threshold voltage. Therefore, it is recommended

TABLE 10.1
Parameter Specifications

Parameters	CGAA FET	MC-DD-Ge/Si-INW-FET
Channel doping	$1 \times 10^{16}/cm^3$	$1 \times 10^{16}/cm^3$
Silicon thickness (t_{si})	10 nm	10 nm
Channel length (L_{ch})	30 nm	30 nm
Oxide thickness (t_{ox})	2 nm	2 nm
Work function (ϕ_m)	4.88 eV	4.6 eV
Gate thickness (t_g)	2 nm	2 nm
Filler thickness (t_f)	5 nm	5 nm

to prioritize a lower r_2/r_1 ratio while ensuring moderate values for both r_1 and r_2 to ensure an adequate number of mobile charges in the channel. In the proposed macaroni structure (MC-DD-Ge/Si-INW-FET), the r_2/r_1 ratio is set at 5 nm, which is moderately low and is expected to help in minimizing fluctuations in the threshold voltage. Table 10.1 provides a synopsis of the parameters comparing conventional gate-all-around FET (CGAA FET) and MC-DD-Ge/Si-INW-FET. The drawback of this configuration lies in its expensive and intricate fabrication process. Challenges include achieving and sustaining precise temperatures for vacuum dielectrics, difficulty in accurate doping concentrations, and the high cost of required fabrication equipment.

Numerical simulations are conducted using an ATLAS 3D device simulator [20], incorporating various models such as the SRH, CONMOB, FLDMOB, hot electron injection (HEI), Concannon nonlocal gate current (N. CONCAN), bandgap narrowing (BBT.STD) [21], and drift diffusion models. The syntax of the band-to-band tunneling (BTBT) BBT.STD model is specifically employed to accurately analyze and implement our proposed DDM-GAA FET device, focusing on gate-induced drain leakage (GIDL) characteristics similar to fabricated devices [22]. Detailed descriptions of these models can be found in Table 10.2. Carrier transport problems are solved numerically using the Newton-Gummel technique [23].

10.3 RESULT AND DISCUSSION

10.3.1 MC-DD-Ge/Si-INW-FET for Analog Applications

Figure 10.2 illustrates the variations in band energy across the channel length for both MC-DD-Ge/Si-INW-FET and CGAA FET under different drain voltages (V_{DS}= 0.0 V and 1.0 V, respectively). In the case of MC-DD-Ge/Si-INW-FET, there is a noticeable decrease in band energy toward the drain end as depicted in the graph. This reduction in band energy is influenced by two factors: the filler acting as a dielectric and the high-*k* dielectric located at the drain side, which significantly reduces electron tunneling from the valence band to the conduction band. Consequently, BTBT, involving electron transition from the valence band to the conduction

TABLE 10.2
Simulation Employed Models

Models	Details
BBT.STD	Employed to assess the impact of charge carrier tunneling
FLDMOB	Utilized to address the impact of velocity saturation
CONMOB	Employed to consider the influence of MOSFET mobility and concentration
HEI	Utilized to incorporate tunneling carriers contributing to gate current
SRH	Utilized for integrating the impact of carrier recombination
Drift diffusion	Incorporates Boltzmann statistics
N. CONCAN	Utilized to analyze substrate current

band, is minimized. In Figure 10.2b, a steep change in band energy can be noticed on the source side because of the presence of germanium as the source and silicon as channel and drain.

Figure 10.3 displays the variation in GIDL current (I_{DS}) with gate voltage for multiple devices. GIDL, a subthreshold leakage phenomenon caused by BTBT, occurs when the gate voltage (V_{GS}) decreases and the drain becomes more positively biased. Electrons emitted due to this phenomenon travel toward the drain end, constituting GIDL current. Figure 10.3 distinctly shows that I_{DS} decreases from 10^{-9}A to 10^{-10}A in the MC-DD-Ge/Si-INW-FET device. This reduction in OFF-state leakage current is primarily attributed to the presence of hot carriers in the filler at the drain side and the high-*k* dielectric situated there.

Figure 10.4 displays the drain current (I_{DS}) characteristics for various devices at different gate voltages (V_{GS}). The graph illustrates that at $V_{GS} = 1.0$ V, the MC-DD-Ge/Si-INW-FET shows superior I_{DS} characteristics in the ON state compared to other designs. This enhancement in drain current is attributed to the dual metal gate and the vacuum gate dielectric employed in the MC-DD-Ge/Si-INW-FET device.

Figure 10.5 illustrates the I_{ON}/I_{OFF} ratio across various device layouts. The I_{ON}/I_{OFF} ratio is a crucial performance metric, as it dictates the efficiency of the device for digital applications. To achieve optimal digital performance, the ratio of current in the ON state to that in the OFF state should be maximized. It can be observed from Figure 10.5 that the MC-DD-Ge/Si-INW-FET exhibits an I_{ON}/I_{OFF} ratio 28 times higher than that of the CGAA FET. This higher ratio in the MC-DD-Ge/Si-INW-FET is due to its superior ON-state current and lower OFF-state current. Therefore, the MC-DD-Ge/Si-INW-FET is well suited for digital applications.

Figure 10.6 presents the subthreshold slope (SS) for several devices. The subthreshold slope characterizes a device's ability to transition between OFF and ON states [24]. Ideally, this slope should be 60 mV/decade for optimal performance. According to Figure 10.6, the CGAA FET exhibits a subthreshold slope of 77 mV/decade, which exceeds the ideal value. On the other hand, the MC-DD-Ge/Si-INW-FET shows a subthreshold slope of 65 mV/decade, slightly above the optimal value but lower compared to the CGAA FET. The presence of a high-*k* dielectric at the

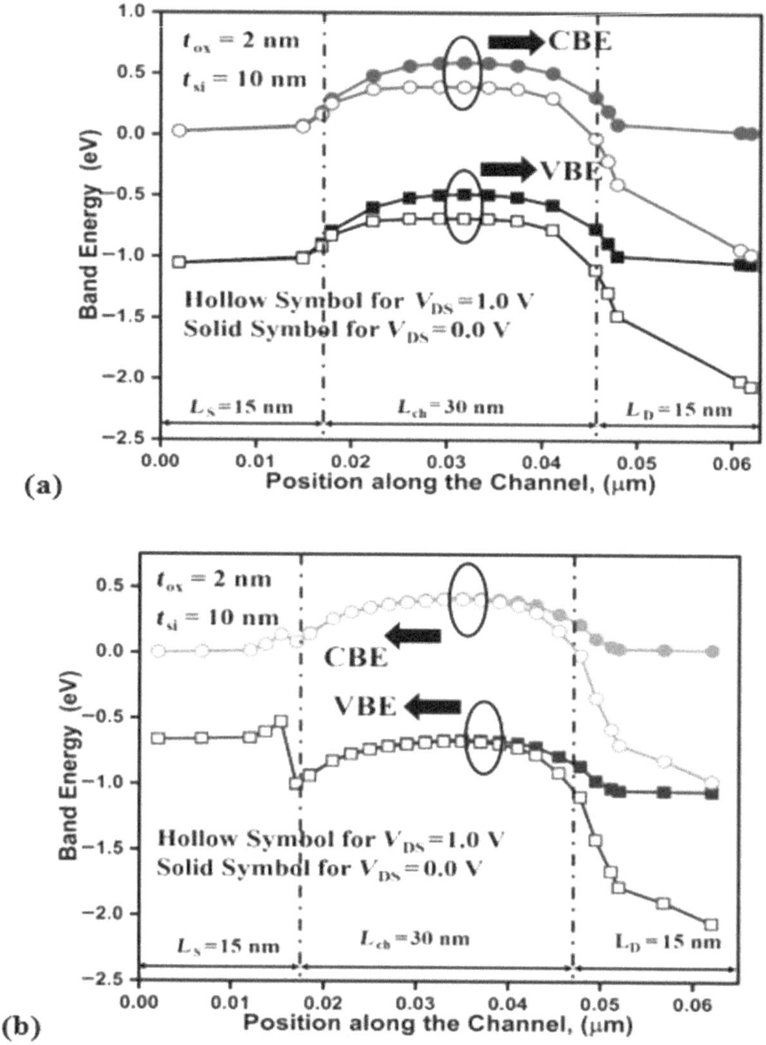

FIGURE 10.2 Energy-band diagram for (a) for CGAA FET and (b) MC-DD-Ge/Si-INW-FET.

drain end and a filler in the MC-DD-Ge/Si-INW-FET significantly reduce the impact ionization effect, thereby improving the transition from the OFF state to the ON state and consequently lowering the subthreshold slope.

Figure 10.7 illustrates the gate voltage–dependent variation in drain current, which is termed "transconductance." The mathematical expression for transconductance is defined by Equation 10.1. The results are compared between CGAA FET and MC-DD-Ge/Si-INW-FET in Figure 10.7. Transconductance is critical not only for analog/RF applications but also for determining the optimal bias point where the

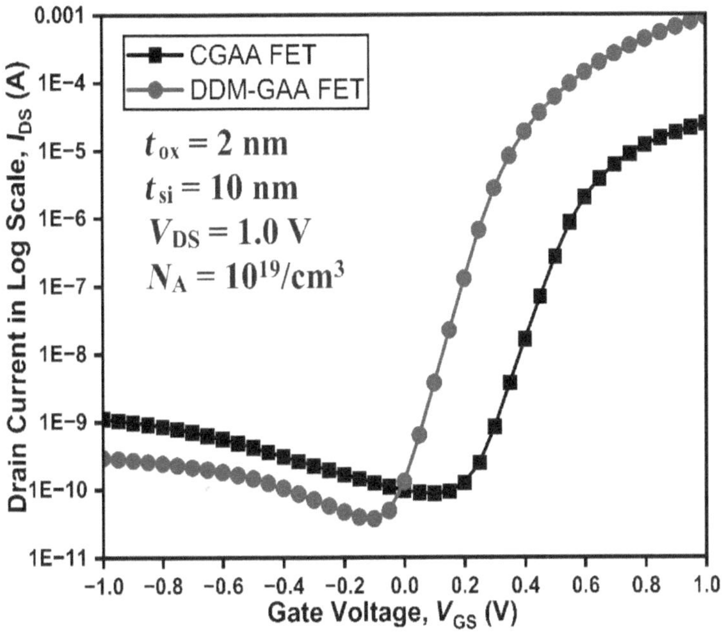

FIGURE 10.3 Off-state leakage along channel for various devices.

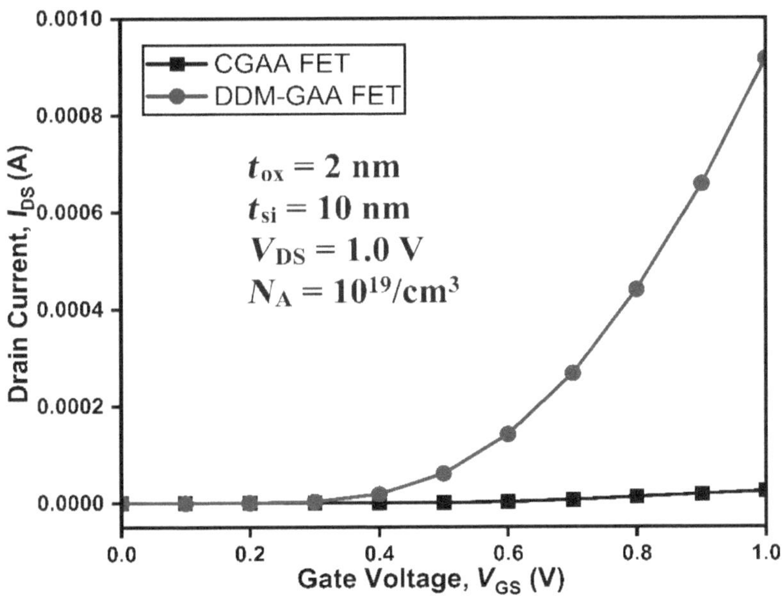

FIGURE 10.4 The variation of drain current with respect to gate voltage.

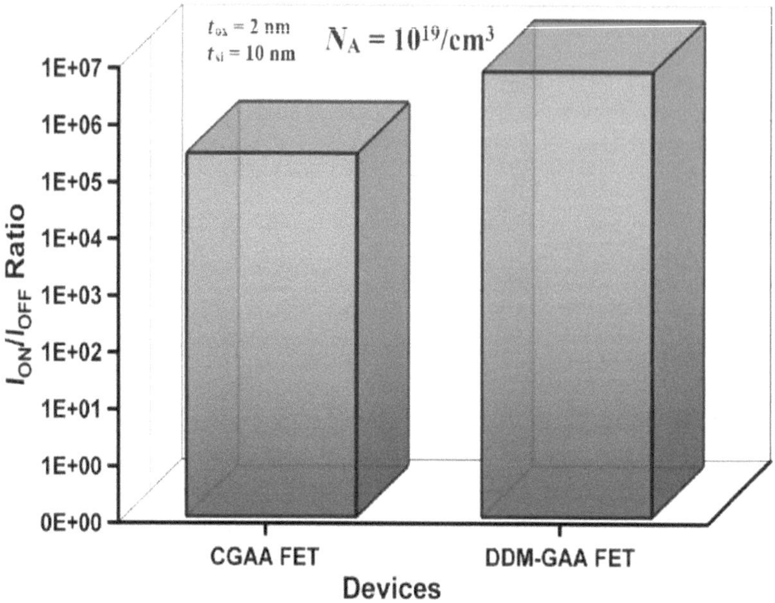

FIGURE 10.5 The variation of I_{ON}/I_{OFF} in various devices.

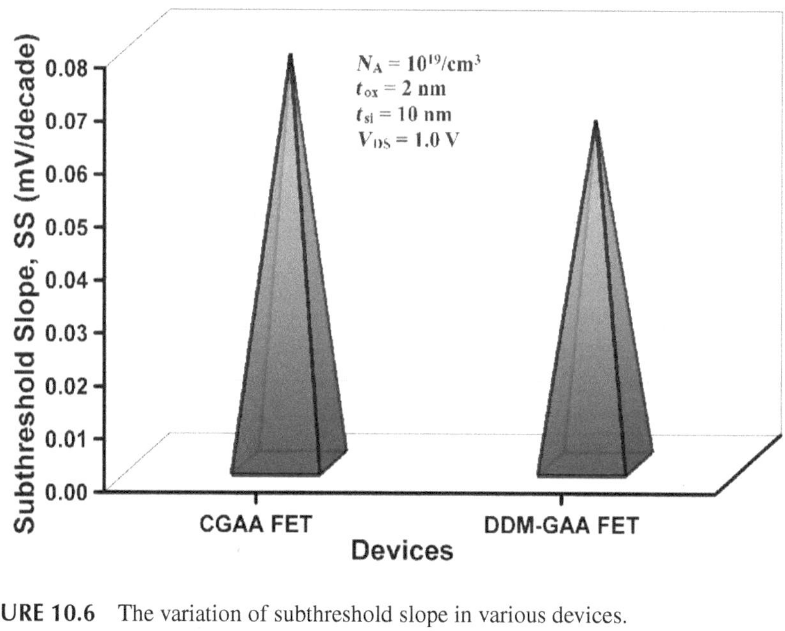

FIGURE 10.6 The variation of subthreshold slope in various devices.

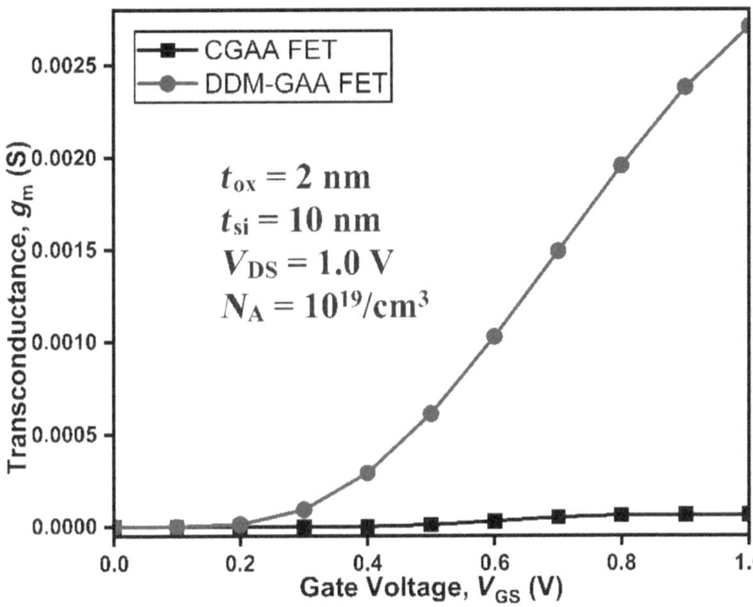

FIGURE 10.7 The variation of transconductance in various devices.

cutoff frequency of each device is minimized. The enhanced current derivability and higher drain current contribute to a higher transconductance value in MC-DD-Ge/Si-INW-FET.

$$g_m = \left.\frac{\partial I_{DS}}{\partial V_{GS}}\right|_{V_{DS}} \tag{10.1}$$

Figure 10.8 displays the output characteristics (I_{DS} vs. V_{DS}) of MC-DD-Ge/Si-INW-FET measured at V_{GS} = 1.0 V for all devices. In this study, the output characteristics of MC-DD-Ge/Si-INW-FET are superior compared to other device topologies. This improvement is attributed to the presence of high-k dielectric and vacuum filler at the center of the device, which mitigates the hot electron effect. Figure 10.9 illustrates the variation of output conductance (g_d) with V_{DS} for CGAA FET and MC-DD-Ge/Si-INW-FET at V_{GS} = 1.0 V. Output conductance (g_d) represents the change in I_{DS} with respect to the change in V_{DS}. The figure indicates that MC-DD-Ge/Si-INW-FET exhibits profiles that align more closely with the optimal characteristics. MC-DD-Ge/Si-INW-FET features higher output conductance and drain current due to the filler and high-k dielectric. This can be mathematically described in Equation 10.2.

$$g_d = \left.\frac{\partial I_{DS}}{\partial V_{DS}}\right|_{V_{GS}} \tag{10.2}$$

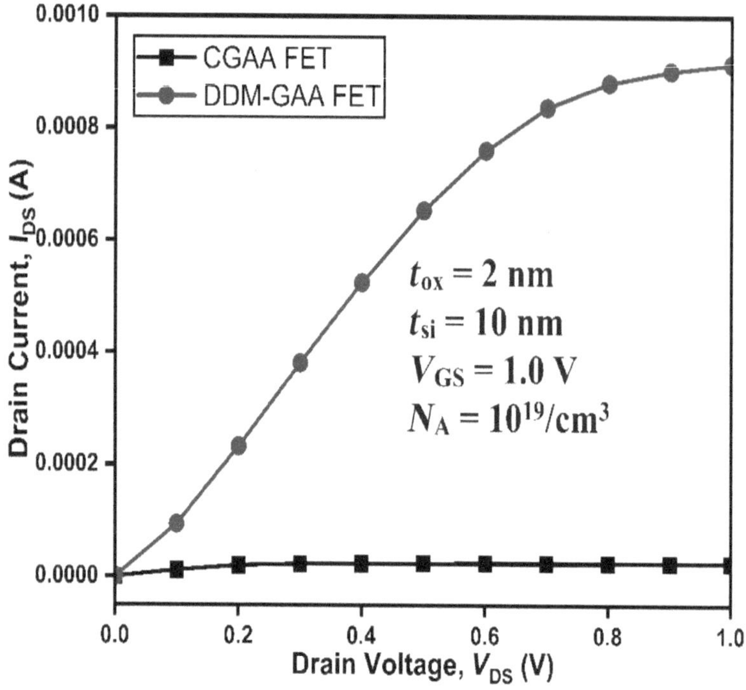

FIGURE 10.8 The variation of subthreshold slope in various devices.

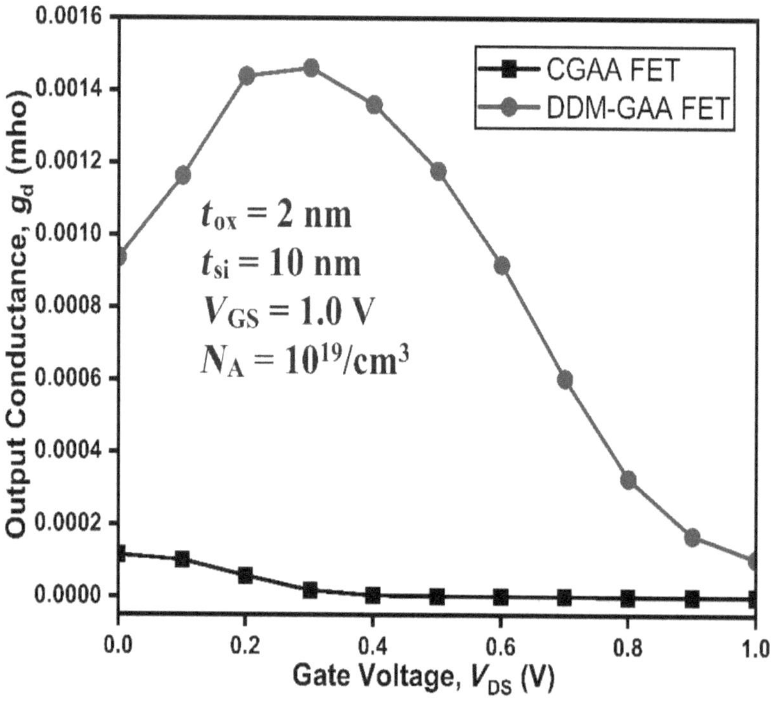

FIGURE 10.9 The variation of output conductance in various devices.

10.3.2 Machine Learning in Enhancing the Analog Performance

FETs are essential elements in analog circuits, serving crucial functions in amplifiers and mixed-signal systems. Conventional approaches to optimizing FET performance typically rely on empirical adjustments and iterative design methods, which can be both time consuming and suboptimal. The introduction of ML techniques offers a promising opportunity to substantially improve the performance and efficiency of FETs in analog applications. This section explores how ML can enhance FET performance by advancing device modeling, optimizing parameters with precision, detecting faults effectively, and improving reliability.

Machine Learning in Device Modeling: ML greatly improves device modeling by offering precise predictions of MOSFET behavior, which is crucial for designing high-performance analog circuits. Unlike traditional models such as SPICE, which use simplified equations, ML harnesses extensive datasets to capture detailed device behavior across various conditions. The process starts with gathering and preprocessing experimental and simulation data, including parameters like threshold voltage, transconductance, and drain current. Supervised learning algorithms, such as support vector machines (SVM), decision trees, and neural networks, are selected based on the complexity of the data. These models are trained and validated using techniques like cross-validation to ensure their effectiveness on new data, with performance assessed using metrics such as mean squared error (MSE) and R-squared (R^2). Ultimately, ML-based models are incorporated into electronic design automation (EDA) tools, enabling engineers to accurately predict key performance metrics and achieve precise control over MOSFET characteristics.

Parameter Optimization: Achieving specific performance goals in MOSFET design through parameter optimization is a significant challenge. Conventional approaches often depend on extensive simulations and heuristic methods, which can be both inefficient and yield less-than-ideal outcomes. ML techniques offer powerful tools for exploring the parameter space more effectively. The process starts with setting objectives and constraints, such as gain, bandwidth, noise figure, and power consumption, while also accounting for manufacturing limitations and device reliability. Optimization algorithms like genetic algorithms (GA), particle swarm optimization (PSO), and Bayesian optimization autonomously search for the best combination of design parameters, including channel length, width, doping concentration, and oxide thickness. These algorithms iteratively assess potential solutions against the specified objectives and constraints, utilizing ML models to predict performance and reduce reliance on time-consuming simulations. Practical applications, such as designing low-noise amplifiers (LNAs), have shown that ML-driven optimization can achieve enhanced performance, offering better noise characteristics and linearity compared to traditional methods.

Fault Detection and Reliability Enhancement: Reliability is a vital factor in MOSFET performance, particularly in analog applications where device failures can have serious implications. ML can significantly improve MOSFET reliability by detecting potential faults and predicting how the device will degrade over time.

Anomaly detection methods, such as autoencoders and one-class SVMs, are trained on data from functioning devices to establish what constitutes normal operation and identify deviations that could signal impending faults. Predictive maintenance models, utilizing approaches like recurrent neural networks (RNNs) and long short-term memory (LSTM) networks, estimate the remaining useful life (RUL) of MOSFETs based on past usage and stress conditions. These ML models are incorporated into monitoring systems that continuously observe MOSFET performance in real time. Early detection of anomalies through these models enables proactive maintenance, thereby minimizing the risk of severe failures. Case studies, including those on power amplifiers (PAs), have shown that ML-based fault detection effectively tracks performance issues caused by thermal stress and aging, resulting in timely maintenance and prolonged device life.

Case Studies and Practical Implementations: Numerous case studies highlight how ML can significantly boost MOSFET performance in analog applications. For example, neural networks have been used to model and refine LNAs, leading to notable gains in noise performance and linearity. Similarly, ML-driven optimization methods have enabled the creation of high-efficiency PAs with improved gain and lower power consumption. ML has also been applied to enhance operational amplifiers, resulting in better bandwidth, offset voltage, and power efficiency. In mixed-signal systems, ML techniques have been utilized to effectively manage the trade-offs between analog and digital performance metrics, producing more efficient and reliable systems.

ML has markedly advanced the optimization and reliability of MOSFETs in analog applications. Unlike traditional methods that depend on extensive simulations and heuristic approaches, ML uses large datasets to build highly accurate and predictive models.

In addition to modeling, ML is crucial for optimizing device parameters and enhancing reliability. Optimization algorithms such as GA, PSO, and Bayesian optimization explore the parameter space efficiently, reducing the need for lengthy simulations. ML also plays a key role in reliability enhancement through anomaly detection algorithms like autoencoders and one-class SVMs, which identify deviations from normal operating conditions. Predictive maintenance models using RNNs and LSTM networks estimate the RUL of MOSFETs, allowing for timely maintenance and minimizing failure risks. Case studies involving LNAs and PAs illustrate that ML-driven optimization improves performance, enhancing noise reduction, linearity, efficiency, and overall device lifespan. Similarly, ML algorithms have optimized operational amplifiers (Op-Amps) to achieve better bandwidth, offset voltage, and power efficiency. In mixed-signal systems, ML helps balance analog and digital performance metrics, resulting in more efficient and reliable systems. These examples demonstrate the transformative potential of ML in MOSFET design, providing robust solutions to complex challenges and driving progress in analog circuit performance.

As the field advances, the integration of ML into MOSFET design processes is anticipated to grow, leading to further innovations and improvements in analog electronics.

10.3.3 MC-DD-Ge/Si-INW-FET for Biosensing Applications

FETs are a great choice for designing biosensors in rapid and high sensitivity–based biosensing applications. The MC-DD-Ge/Si-INW-FET is realized as a biosensor by etching a dielectric layer from the source and drain side. The dual-sided cavities created are utilized for biomolecule immobilization. Figure 10.10 depicts the 2D view of the structure of the MC-DD-Ge/Si-INW-FET based biosensor device (showing cavity). The two-sided nanocavity has been formed by etching the SiO_2 layer on the source side and the HfO_2 dielectric layer on the drain side. The thickness of the nanocavities (t_{cavity}) is 6 nm, and the length of the nanocavities (L_{cavity}) is 10 nm each.

By immobilizing a variety of biomolecules inside the nanocavity, this biosensor aims to characterize their behavior in a biosensing environment. These biomolecules can be identified by their dielectric constant values: APTES ($K = 3.57$) [25], hydroprotein ($K = 5$) [26], keratin ($K = 8$) [27], gelatin ($K = 12$) [28], and DNA [10, 28] which is distinguished by its charge density values ($N_f = -1 \times 10^{11}/cm^2$, $N_f = -5 \times 10^{11}/cm^2$, $N_f = -1 \times 10^{12}/cm^2$) and dielectric constant value $K = 12$. When distinct

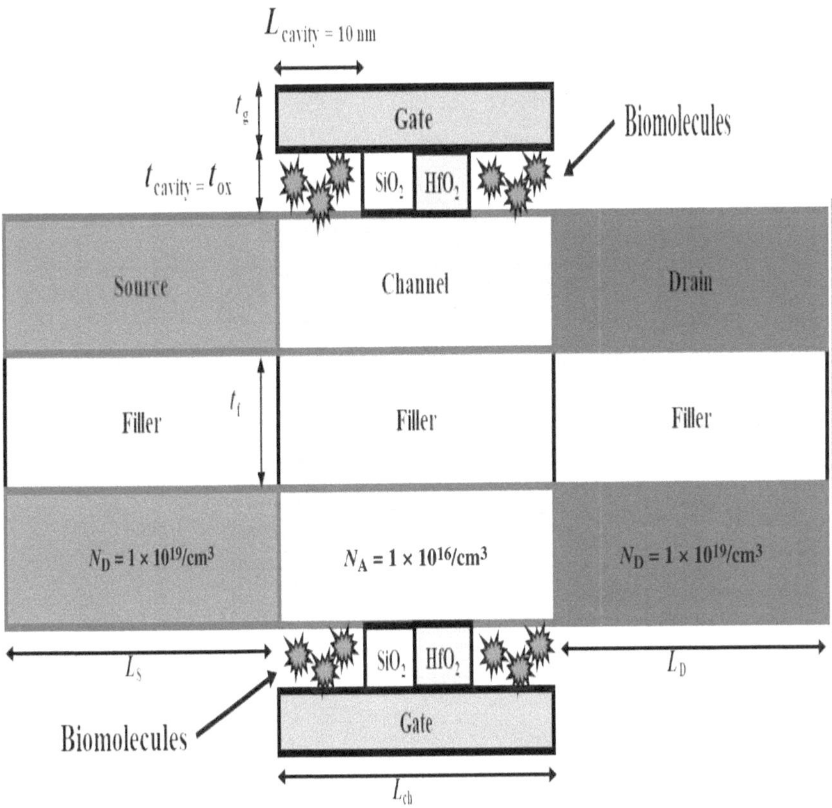

FIGURE 10.10 Two-dimensional structure of MC-DD-Ge/Si-INW-FET as a biosensor.

biomolecules are immobilized in FET-based biosensors, the gate oxide capacitance of the device changes, affecting the potential profile, electric field distribution, and I_D as a function of both V_{GS} and V_{DS}. Changes in these electrical properties are frequently employed in identifying particular biomolecules.

Figure 10.11 illustrates the shift in surface potential caused by the immobilization of different neutral biomolecules and DNA biomolecules with different negative charges inside the nanocavity. The surface potential drops from APTES to gelatin, as shown in Figure 10.11a. Region 1 in the channel beneath the source-side nanocavity exhibits the largest change in surface potential for various biomolecules compared to the absence of any biomolecule. This results from the drain voltage decreasing monotonically as it approaches the source. The charged DNA biomolecules with the lowest surface potential among those studied have a charge density of $N_f = -1 \times$

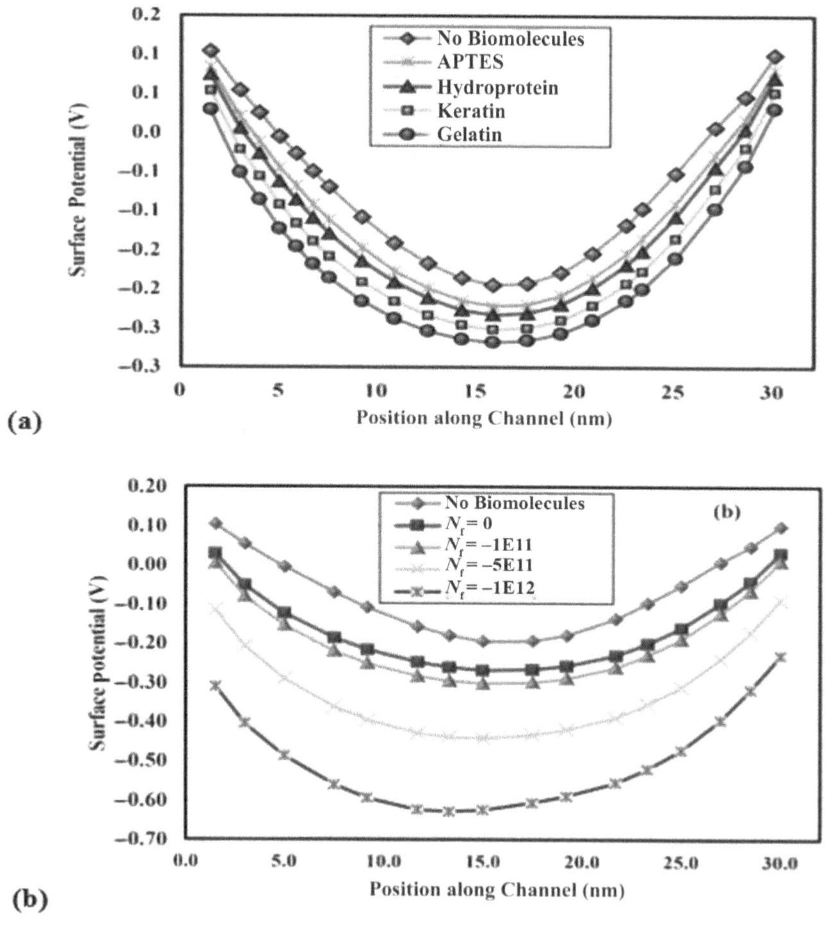

FIGURE 10.11 Surface potential along the channel of MC-DD-Ge/Si-INW-FET biosensor for (a) various neutral biomolecules and (b) DNA biomolecules at $K = 12$ and varying charges.

$10^{12}/cm^2$, as shown in Figure 10.11b. For $N_f = -1 \times 10^{12}/cm^2$, we thereby achieve the biggest changes in surface potential and, thus, maximal sensitivity out of all DNA biomolecules trapped in the nanocavity region. A decrease in potential results from the rise in flat band voltage brought on by qN_f/C_{eff}. Consequently, this means that there is more channel area depletion [26].

The change in the MC-DD-Ge/Si-INW-FET biosensor's electric field for charged and neutral biomolecules is shown in Figure 10.12. The electric field for biomolecules with different dielectric constant values, from APTES to gelatin, is displayed along the channel length in Figure 10.12a. It is evident that the electric field at the source side rises as the dielectric constant increases [21, 23, 29].

An electric field plot of negatively charged DNA biomolecules vs. neutral biomolecules at a dielectric constant $K = 12$ is shown in Figure 10.12b, along with a scenario without any biomolecules. The graph unequivocally demonstrates that the electric field varies noticeably when DNA biomolecules are present. The electric field strength

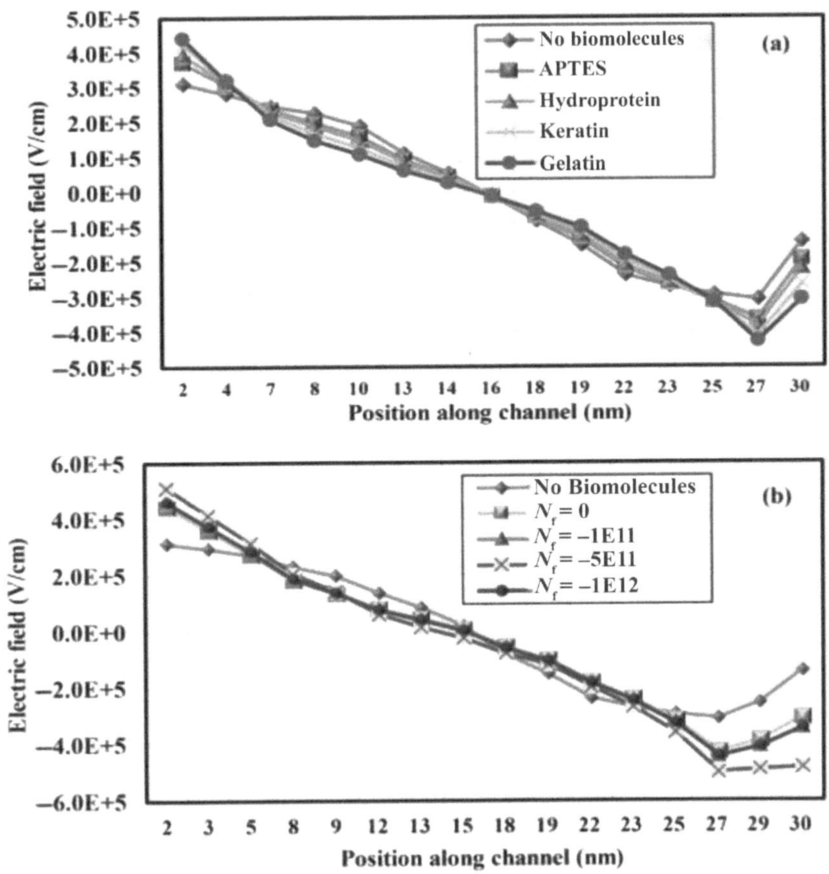

FIGURE 10.12 Electric field of MC-DD-Ge/Si-INW-FET biosensor for (a) various neutral biomolecules and (b) DNA biomolecules at $K = 12$ and varying charges.

increases near the source region as the DNA biomolecules' negative charge intensifies. However, as it approaches the drain, the electric field in the channel gradually decreases. Furthermore, when the negative charge of the DNA grows, Figure 10.12b shows that the electric field is reduced to its lowest point at the channel-drain junction. Essentially, the changed characteristics of the gate electrode and its interaction with the DNA biomolecules are responsible for this shift in the electric field.

The drain ON-current characteristics for immobilizing several biomolecules, including gelatin, keratin, hydroprotein, and APTES, are shown in Figure 10.13a, whereas the drain current on a logarithmic scale representing the subthreshold current $V_{GS} = 0.0$ V is shown in Figure 10.13b. The ON current rises, and the subthreshold

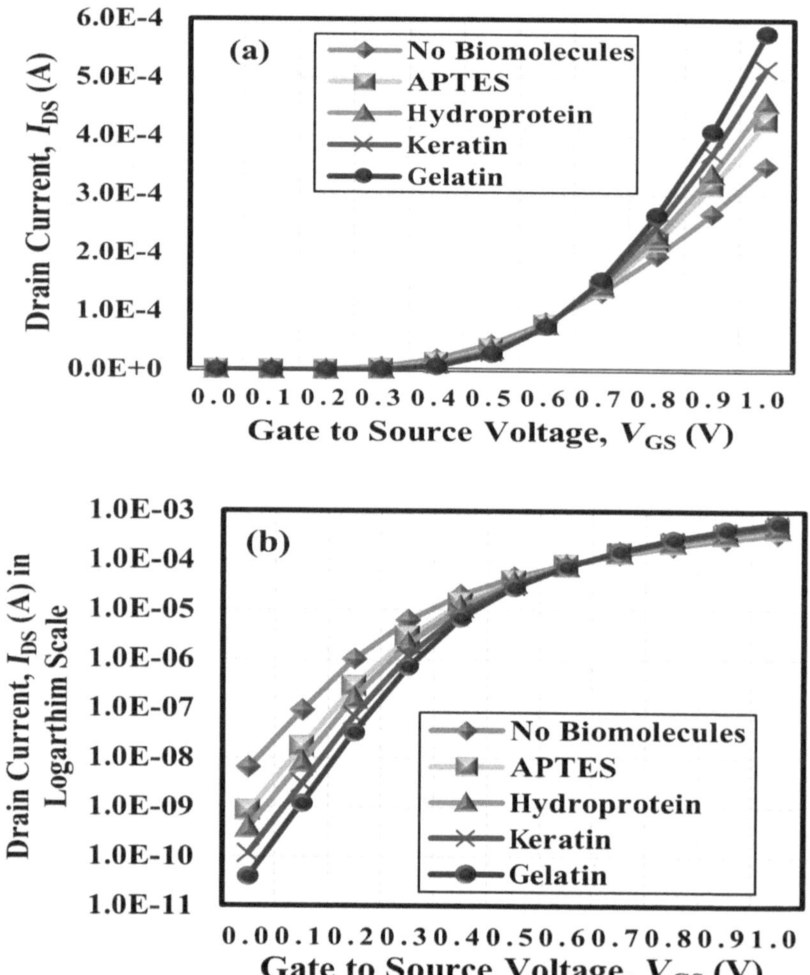

FIGURE 10.13 (a) I_D–V_{GS} characteristics with different biomolecules with changing dielectric constant, (b) drain current in logarithmic scale representing the subthreshold current at $V_{GS} = 0.0$ V, and (c) transconductance for different biomolecules.

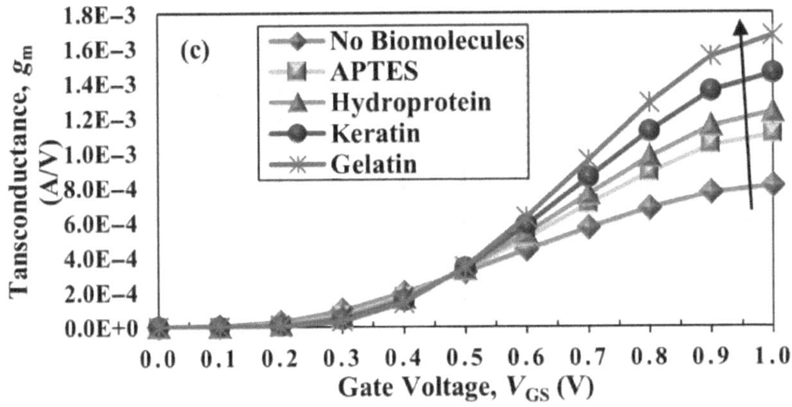

FIGURE 10.13 (Continued)

current falls as the K values of various biomolecules rise. To identify certain biomolecules, this change in current is an essential sensing parameter.

The proposed device's transconductance (g_m) is illustrated in Figure 10.13c upon the introduction of several neutral biomolecules into the nanocavity region. Transconductance (g_m) is the first-order derivative of drain current with respect to gate voltage (V_{GS}). Figure 10.13c shows that the change in transconductance (g_m) increases progressively when different biomolecules are taken into account, ranging from APTES to gelatin. With a transconductance (g_m) peak value of 1.67×10^3 μA/V, the gelatin biomolecule has the greatest value. This result emphasizes that when the dielectric constant value of the biomolecules increases, the device's capacity to transform tiny input voltage perturbations into corresponding variances in current is maximized.

The drain current characteristics for charged DNA biomolecules are displayed in Figure 10.14, in which Figure 10.14a shows the drain ON current while Figure 10.14b displays the OFF current on a logarithmic scale. When biomolecules are negatively charged, the drain ON as well as OFF current values drop [28]. A discernible decrease in drain current is observed in the suggested MC-DD-Ge/Si-INW-FET biosensor. The impact of the gate increases with the introduction of more negatively charged proteins, which causes a higher degree of carrier depletion and, hence, lower amounts of OFF current. For example, in the case of $K = 12$ and $N_f = -1 \times 10^{12}/\text{cm}^2$, the OFF current is 1.47×10^{-14} A, whereas in the case of no biomolecule, it is 3.99×10^{-11} A.

Figure 10.15 shows the drain current in relation to the drain to source voltage (V_{DS}). The variation in drain current for altering the dielectric constant of biomolecules is shown in Figure 10.15a, and the variation in drain ON current, I_{DS}(ON), for altering the negative charge of DNA biomolecules is demonstrated in Figure 10.15b. When the dielectric constant remains constant, and the negative charge of DNA rises, the drain ON current shows a declining trend [30–32]. The decrease in drain current is caused by a downward shift in surface potential when negatively charged biomolecules are immobilized [26].

The most important component in sensor design is sensitivity. The conventional strategy for testing the sensitivity of a FET biosensor is to analyze the proportional change in

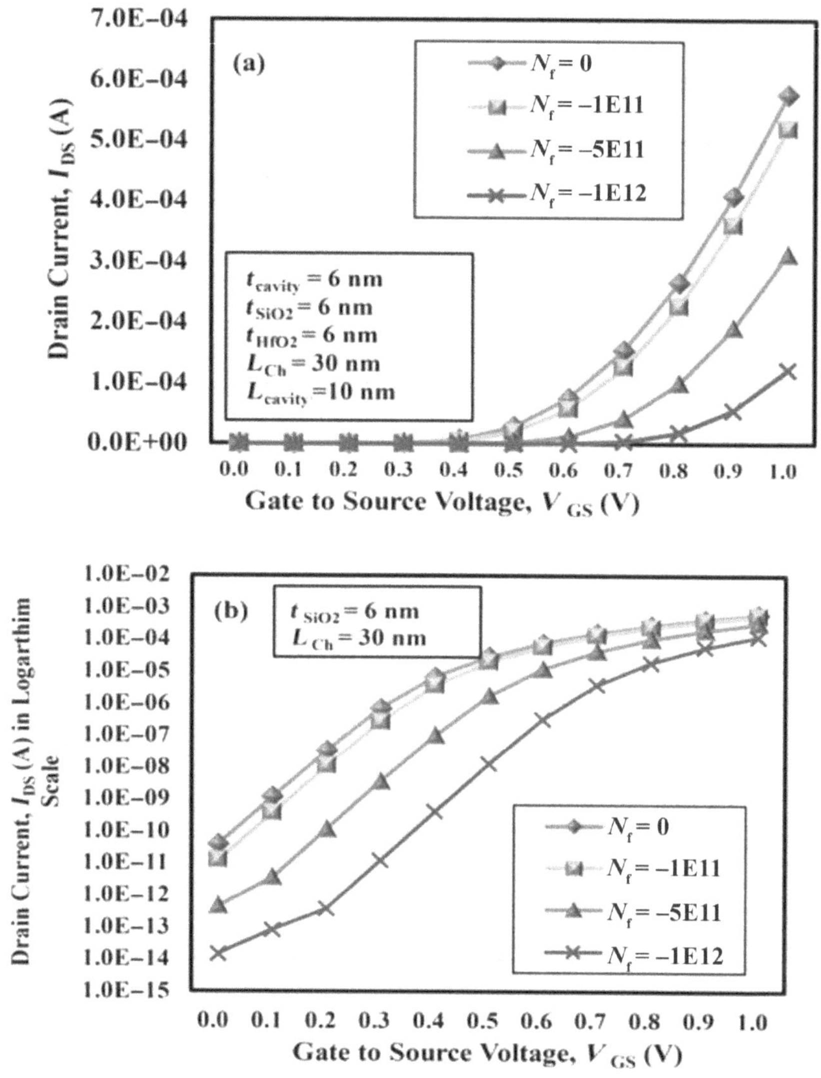

FIGURE 10.14 (a) I_D–V_{GS} characteristics of MC-DD-Ge/Si-INW-FET biosensor for charged DNA biomolecules and (b) drain current in logarithmic scale for charged DNA biomolecules.

a quantifiable electrical parameter that reflects the presence of biomolecules [10, 11, 33, 34]. This electrical characteristic can include I_{ON} current and V_{th}, among other parameters. The formal definition of sensitivity (S_{EP}) in mathematics is as follows:

$$S_{EP} = \left| \text{EP}_{\text{No Biomolecule}} \text{-} \text{EP}_{\text{Biomolecules}} \right| \tag{10.3}$$

The electrical parameter value before being exposed to the biomolecules is denoted as $\text{EP}_{\text{No Biomolecule}}$, and the electrical parameter value of the biosensor after the biomolecules are immobilized is $\text{EP}_{\text{Biomolecules}}$.

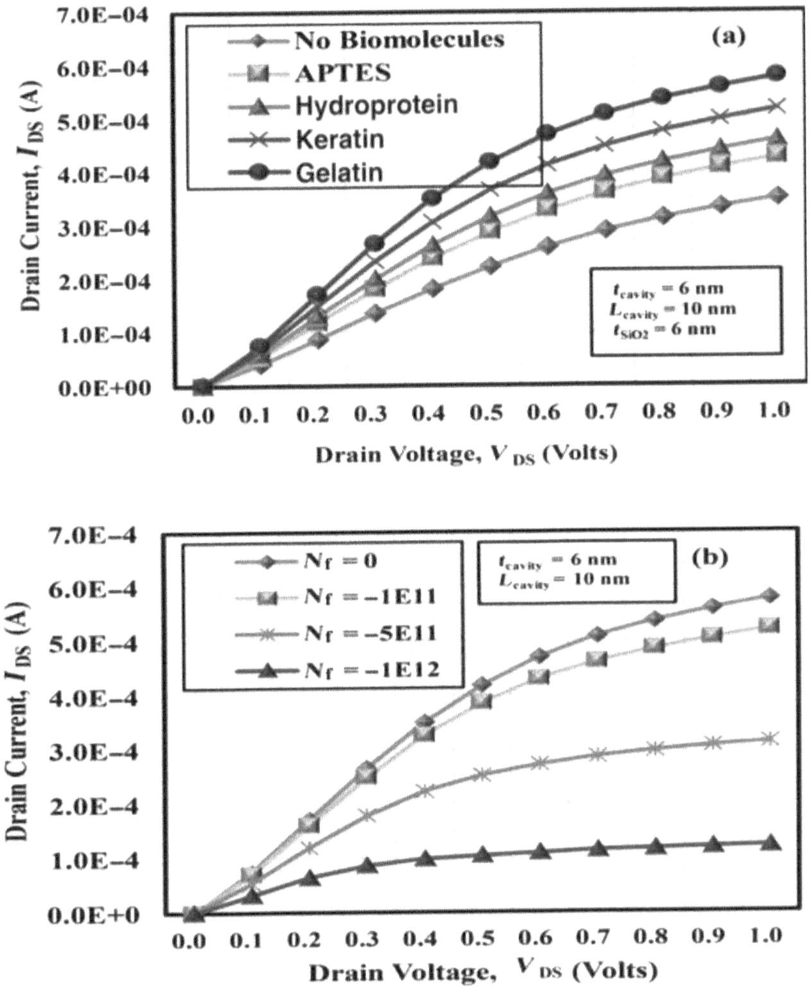

FIGURE 10.15 I_D–V_{DS} characteristics of MC-DD-Ge/Si-INW-FET biosensor for (a) neutral biomolecules and (b) charged DNA biomolecules.

The MC-DD-Ge/Si-INW-FET for neutral biomolecules' drain current sensitivity is shown in Figure 10.16a. Equation 10.3 yielded the following results, which are shown in Figure 10.16a: drain current sensitivity is systematically increased across all gate voltage values upon the addition of biomolecules, from APTES to gelatin. This pattern indicates that biomolecules with higher dielectric constants are more sensitive.

For DNA biomolecules with varying charge concentrations, the drain current sensitivity is shown in Figure 10.16b. When compared to neutral biomolecules, negatively charged biomolecules have a higher binding capacity. As a result, the flow of charge carriers across the channel is more tightly regulated when negatively charged biomolecules are present, which raises the sensitivity of the drain current.

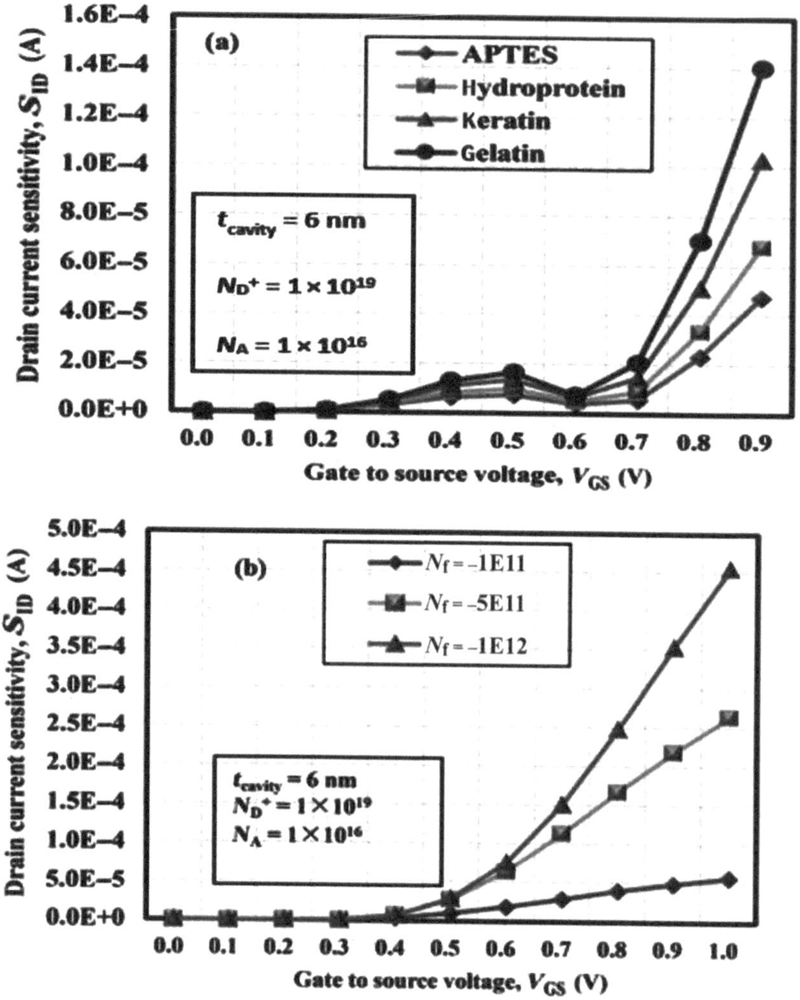

FIGURE 10.16 Drain current sensitivity for (a) different biomolecules and (b) DNA biomolecules with increased negative charge.

For $K = 12$ and charge concentration $N_f = -1 \times 10^{12}/\text{cm}^2$, the drain current sensitivity is 455 μA.

In FET-based biosensors, the variation in threshold voltage is crucial for assessing the device's sensitivity. When biomolecule permittivity increases, particularly in the presence of neutral and negatively charged DNA biomolecules at $K = 8$, the threshold voltage rises, as illustrated in Figure 10.17a and 10.17b. This occurs because a higher gate voltage is required to fully deplete the channel when the surface potential of the channel decreases from the absence of biomolecules to the presence of neutral and negatively charged DNA biomolecules. Consequently, this leads to an increase

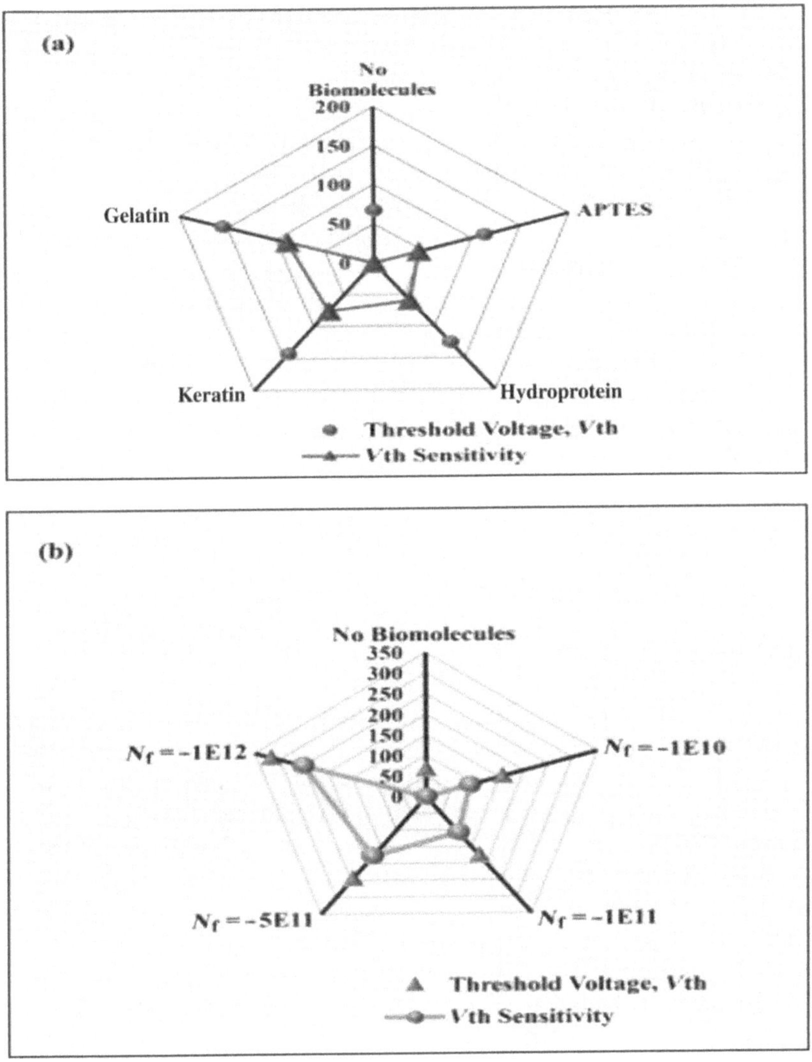

FIGURE 10.17 Threshold voltage and threshold voltage sensitivity for various (a) neutral biomolecules and (b) charged DNA biomolecules.

in the threshold voltage. The need for a greater gate voltage to deplete the channel in the presence of biomolecules indicates that the channel's surface potential is significantly affected by the biomolecules' permittivity and charge.

The gate oxide capacitance rises with the increasing dielectric constant of biomolecules [9], which significantly contributes to the relative change in threshold voltage and thus the sensitivity. For APTES and keratin biomolecules, the V_{th} sensitivity is 45.8 mV and 75.4 mV, respectively. In contrast, for DNA biomolecules with $N_f = -1 \times 10^{11}/\text{cm}^2$ and $N_f = -1 \times 10^{12}/\text{cm}^2$, the V_{th} sensitivity are 107.14 mV and 249.50 mV, respectively, as depicted in Figure 10.17b. The increase in threshold voltage sensitivity with higher K and N_f values demonstrates the enhanced response of the biosensor to variations in biomolecule properties, making it more effective in detecting different biomolecular interactions.

Figure 10.18 displays the I_{ON}/I_{OFF} ratio sensitivity for different biomolecules. The ratio I_{ON}/I_{OFF} for several biomolecules is shown in the inset of Figure 10.18. The gate oxide capacitance increases with the biomolecule's dielectric constant, providing the gate more control over charge carriers and raising the ON current. Subthreshold current is the term for the current that exists even before it reaches the threshold voltage (V_{th}) [35]. Reduced drain OFF current results from the channel being weakly inverted early by the higher oxide capacitance [33]. The dielectric constant K of the biomolecules increases as a result, improving the I_{ON}/I_{OFF} ratio. Because of its larger fluctuations in value for minute changes in the dielectric constant, it is worth mentioning that the I_{ON}/I_{OFF} ratio can also be used as a metric to assess the sensitivity of the biosensor. Equation 10.4 was used to calculate the sensitivity of the I_{ON}/I_{OFF} ratio [36]:

$$\frac{I_{ON}}{I_{OFF}}\text{ Sensitivity : } S_{I_{ON}/I_{OFF}} = \frac{\left.\frac{I_{ON}}{I_{OFF}}\right|_{\text{With biomolecules}}}{\left.\frac{I_{ON}}{I_{OFF}}\right|_{\text{No biomolecules}}} \tag{10.4}$$

A comprehensive sensitivity analysis was conducted to evaluate the performance of the DDM-GAA FET biosensor in comparison to other FET biosensors with similar structural characteristics. In Figure 10.19, it is evident that the proposed biosensor exhibits notably higher S_{Vth} when compared to several recent biosensors, including the GaN-GME-DE-SNW-FET [17], GC-GAA-NWFET [37], DG-6H-SiC SB-FET [36], and DM DPDG-TFET [38], for $N_f = -1E12/cm^2$. The MC-DD-Ge/Si-INW-FET biosensor's great sensitivity highlights its potential for improved detection capabilities in biosensing applications, especially for DNA biomolecules. The advantages of the macaroni channel and dual dielectric structure place the MC-DD-Ge/Si-INW-FET biosensor in a competitive and useful position in the biosensing technology sector.

10.3.4 Machine Learning in Improving the Performance of Biosensor

ML has significantly enhanced FET-based biosensors by improving their sensitivity and specificity, making them more effective in medical diagnostics, environmental monitoring, and biotechnology. BioFETs, sophisticated biochemical sensors operating like FETs, now benefit from real-time detection of molecules like proteins, DNA, and pathogens [19]. Training ML models on large datasets allows for accurate detection with lower false positive rates. This integration enables researchers to model optimal sensor designs before fabrication, speeding up development and reducing costs. ML refines sensor parameters over time, adjusting to new biological targets and analyzing complex data with models such as neural networks. This results in more precise sensor responses, which are essential for early

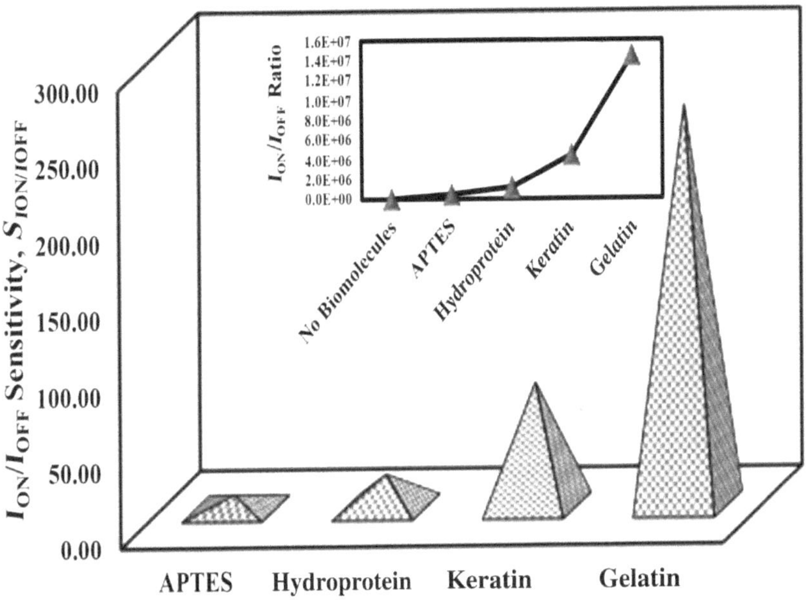

FIGURE 10.18 I_{ON}/I_{OFF} ratio and I_{ON}/I_{OFF} sensitivity of DDM-GAA FET biosensor.

disease detection and personalized treatment. Additionally, ML enables real-time monitoring and feedback systems, crucial for dynamic environments like in vivo diagnostics and environmental monitoring. Portable BioFETs incorporating ML algorithms facilitate molecular biosensing outside conventional labs, enhancing point-of-care testing, remote health monitoring, and rapid field diagnostics. This innovation is set to transform health care by making BioFETs more functional, sensitive, and intelligent through advanced data analysis, predictive modeling, and real-time monitoring.

Integrating ML into BioFET development involves several key techniques and considerations. First, data preprocessing prepares raw sensor data for analysis by cleaning, standardizing, and improving accuracy through methods like outlier detection, noise reduction, and normalization. Dimensionality reduction techniques, such as principal component analysis (PCA) are also applied. Second, feature extraction transforms raw data into usable inputs for ML models, with features such as changes in current, voltage, or resistance due to biological interactions being extracted using signal processing methods like Fourier and wavelet transforms. Convolutional neural networks (CNNs) can automate the extraction of complex features, detecting intricate patterns in biological interactions. Third, choosing the appropriate ML model—such as SVMs, neural networks, random forests, or *K*-nearest neighbors—is crucial for accurate analysis, based on the data type and analysis requirements. Fourth, real-time adaptation involves regularly updating models with new data to maintain accuracy despite environmental changes, using techniques like online learning and reinforcement learning. Fifth, ensemble methods enhance performance and

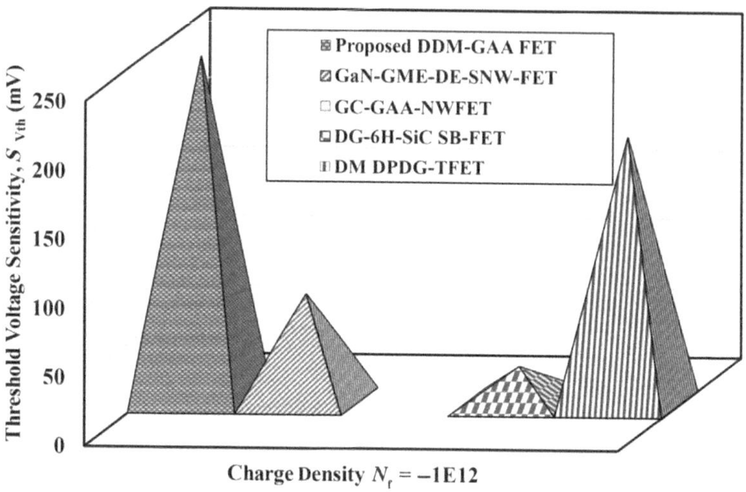

FIGURE 10.19 S_{Vth} comparison of MC-DD-Ge/Si-INW-FET biosensor with recently published FET biosensors.

robustness by combining predictions from multiple models through strategies such as bagging, boosting, and stacking. Sixth, transfer learning improves model accuracy and accelerates deployment by utilizing pretrained models from related tasks, which is particularly beneficial for small datasets. Finally, predictive maintenance and calibration use ML to automate monitoring, optimize conditions, and schedule service for BioFET devices, ensuring precise control and longevity. Overall, ML enhances the sensitivity, specificity, and versatility of BioFETs, making them valuable tools for medical, environmental, and biotechnological applications through advanced preprocessing, feature extraction, ensemble methods, transfer learning, and real-time adaptive model selection.

Data Analysis Based on ML: In data analysis, PCA, SVM, and artificial neural networks (ANNs) are three prominent ML methods applied to BioFET sensor data. PCA is a technique used for reducing dimensionality and extracting key features from complex, high-dimensional sensor data, converting it into principal components that capture the essential variability. This process simplifies the data while maintaining important details about biological interactions. SVM models are effective for both classification and regression tasks. They create optimal hyperplanes to classify biological targets in the sensor data and manage nonlinear relationships without overfitting. ANNs, including CNNs and RNNs, are proficient at identifying complex patterns within large datasets, analyzing sensor data to predict interactions and detect patterns in biological systems. These techniques work together, with PCA filtering out noise and creating features for SVM or ANN analysis, SVM classifying datasets, and ANNs refining intricate data relationships. Together, these methods improve the accuracy, speed, and depth of BioFET applications in the fields of biomedicine and biotechnology.

BioFET Design Based on ML: Recently, the design of sensors, especially BioFETs, has been greatly enhanced by the application of ML algorithms. These advancements have transformed sensor development by boosting operational efficiency, upgrading materials, reducing design cycle times, and improving receptor selection. ML significantly improves BioFET performance by optimizing parameters for greater sensitivity, selectivity, and efficiency, as well as refining 2D material properties like semiconducting transition metal dichalcogenide (TMDC), predicting bandgaps, and optimizing defect structures [39]. Techniques such as Gaussian process regression and Bayesian optimization accelerate design processes for heterostructures, while random forest and gradient boosting enhance electron flow in organic FETs [40]. Additionally, ML revolutionizes receptor selection by predicting and optimizing receptors like enzymes and aptamers, leading to the discovery of novel receptors with high specificity and detection efficiency. Incorporating ML into BioFET design represents a groundbreaking approach, facilitating the precise and efficient creation of sensors with improved sensitivity, specificity, and versatility [41]. This positions ML as a pivotal element in advancing health diagnostics, environmental monitoring, and disease detection, highlighting its crucial role in developing sensor technologies and tackling complex challenges across various fields [42, 43].

10.4 CONCLUSION

This study explores the sensitivity analysis of a macaroni channel integrated with a dual-dielectric Ge/Si nanowire FET, evaluating its performance in various analog applications. Comprehensive simulation tests were conducted to assess the device's analog performance, analyzing a range of parameters and characteristics. The findings highlight significant improvements due to the macaroni channel, germanium source, and horizontal gate stack, which enhance the device's performance compared to traditional models. The I_{ON}/I_{OFF} ratio has increased by approximately 28 times, which is crucial for digital applications, while transconductance has improved by about 44 times, beneficial for amplification. Additionally, the device has been employed as a biosensor, showing high sensitivity to a variety of biomolecules. By examining both charged and neutral biomolecules, the study enhances the applicability and accuracy of its results. The biosensor exhibits a voltage sensitivity of about 75.4 mV for $K_{bio} = 8$. Furthermore, the role of ML in optimizing the device's performance and design is explored through a detailed discussion.

REFERENCES

1. K. Upasana, R. Narang, M. Saxena, and M. Gupta, "Modeling and TCAD assessment for gate material and gate dielectric engineered TFET architectures: Circuit-level investigation for digital applications," *IEEE Transactions on Electron Devices*, vol. 62, no. 10, pp. 3348–3356, 2015.
2. A. Sirohi, C. Sahu, and J. Singh, "Analog/RF performance investigation of dopingless FET for ultra-low power applications," *IEEE Access*, vol. 7, pp. 141810–141816, 2019.
3. J. Park, H. Hiep, A. Woubit, and M. Kim, "Applications of field effect transistor (FET) – type biosensors," *Applied Science and Convergence Technology*, vol. 23, no. 2, pp. 61–71, 2014.

4. A. Das, B. K. Kanaujia, V. Nath, S. Rewari, and R. S. Gupta, "Impact of reverse gate oxide stacking on gate all around tunnel FET for high frequency analog and RF applications," in *2020 IEEE 17th India Council International Conference, INDICON 2020*, pp. 1–6, 2020.
5. R. Gautam, M. Saxena, R. S. Gupta, and M. Gupta, "Gate all around nanowire MOSFET with catalytic metal gate for gas sensing applications," *IEEE Transactions on Nanotechnology*, vol. 12, no. 6, pp. 939–944, 2013.
6. S. H. Kim, S. Agarwal, Z. A. Jacobson, P. Matheu, C. Hu, and T. J. K. Liu, "Tunnel field effect transistor with raised germanium source," *IEEE Electron Device Letters*, vol. 31, no. 10, pp. 1107–1109, 2010.
7. P. S. Goley, and M. K. Hudait, "Germanium based field effect transistors: Challenges and opportunities," *Materials (Basel).*, vol. 7, no. 3, pp. 2301–2339, 2014.
8. I. Chahardah Cherik, and S. Mohammadi, "Vertical tunneling field-effect transistor with germanium source and T-shaped silicon channel for switching and biosensing applications: A simulation study," *IEEE Transactions on Electron Devices*, vol. 69, no. 9, pp. 5170–5176, 2022.
9. A. Chakraborty, and A. Sarkar, "Analytical modeling and sensitivity analysis of dielectric modulated junctionless gate stack surrounding gate MOSFET (JLGSSRG) for application as biosensor," *Journal of Computational Electronics*, vol. 16, no. 3, pp. 556–567, 2017.
10. A. Das, S. Rewari, B. K. Kanaujia, S. S. Deswal, and R. S. Gupta, "Analytical investigation of a triple surrounding gate germanium source metal – oxide – semiconductor field-effect transistor with step graded channel for biosensing applications," *The International Journal of Numerical Modelling: Electronic Networks, Devices and Fields*, vol. 36, no. 6, p. e3106, 2023, https://doi.org/10.1002/jnm.3106.
11. A. Das, S. Rewari, B. K. Kanaujia, S. S. Deswal, and R. S. Gupta, "Ge/Si interfaced label free nanowire BIOFET for biomolecules detection – analytical analysis," *Microelectronics Journal*, vol. 138, p. 105832, 2023, https://doi.org/10.1016/j.mejo.2023.105832.
12. S. I. Amin, and R. K. Sarin, "Enhanced analog performance of doping-less dual material and gate stacked architecture of junctionless transistor with high-k spacer," *Applied Physics A - Applied Physics A: Materials Science*, vol. 122, no. 4, p. 380, 2016.
13. A. Das, S. Rewari, B. K. Kanaujia, and R. S. Gupta, "Recent technological advancement in surrounding gate MOSFET for biosensing applications – a synoptic study," *Silicon*, vol. 14, pp. 5133–5143, 2022.
14. A. Goel, S. Rewari, and S. Verma, "Novel dual-metal junctionless nanotube field-effect transistors for improved analog and low-noise applications," *Journal of Electronic Materials*, vol. 50, no. 1, pp. 108–119, 2021, https://doi.org/10.1007/s11664-020-08541-9.
15. P. Goyal, G. Srivastava, J. Madan, R. Pandey, and R. S. Gupta, "Source material valuation of charge plasma based DG-TFET for RFIC applications," *Semiconductor Science and Technology*, vol. 37, no. 9, p. 095023, 2022.
16. A. Das, S. Rewari, B. K. Kanaujia, and R. S. Gupta, "Recent technological advancement in surrounding gate MOSFET for biosensing applications – a synoptic study," *Silicon*, vol. 14, no. 10, pp. 5133–5143, 2022.
17. S. Sharma, V. Nath, S. S. Deswal, and R. S. Gupta, "Analytical modelling and sensitivity analysis of Gallium nitride-gate material and, dielectric engineered- Schottky nano-wire fet(GaN-GME-DE-SNW-fet) based label-free biosensor," *Microelectronics Journal*, vol. 129, p. 105599, June 2022.
18. H. Amrouch, V. M. van Santen, J. Diaz-Fortuny, and F. Klemme, "Machine learning unleashes aging and self-heating effects: From transistors to full processor," in *2024 IEEE International Reliability Physics Symposium (IRPS)*, pp. 1–8, 2024, https://doi.org/10.1109/IRPS48228.2024.10529386.
19. X. Xhafa, A. L. I. D. O. Gu, and M. B. Yelten, "Machine learning-based modeling of hot carrier injection in 40 nm CMOS transistors," *IEEE Journal of the Electron Devices Society.*, vol. 12, pp. 281–288, 2024.

20. "ATLAS User's Manual." 3-D device simulator. SILVACO International, Version5.14.0.R, 2019.
21. A. Kaul, S. Rewari, and D. Nand, "Macaroni channel-nanowire-field effect transistor (MC-NW-FET) for gate induced drain leakage (GIDL) reduction application," in *Proceedings of the 2022 IEEE International Conference of Electron Devices Society Kolkata Chapter, EDKCON 2022*, pp. 35–38, 2022, https://doi.org/10.1109/EDKCON56221.2022.10032964.
22. A. Kaul, S. Rewari, and D. Nand, "Hetero-dielectric macaroni channel cylindrical gate all around field effect transistor (HD-MC CGAA FET) for reduced gate leakage analog applications," *Microsystem Technologies*, 2023, https://doi.org/10.1007/s00542-023-05577-9.
23. A. Kaul, S. Yadav, S. Rewari, and D. Nand, "Computational modelling of cylindrical-ferroelectric-dual metal-nanowire field effect transistor (C-FE-DM-NW FET) using landau equation for gate leakage minimization," *Micro Nanostructures*, vol. 191, 2024, https://doi.org/10.1016/j.micrna.2024.207851.
24. A. Das, B. K. Kanaujia, S. S. Deswal, S. Rewari, and R. S. Gupta, "Doping induced threshold voltage and ION/IOFF ratio modulation in surrounding gate MOSFET for analog applications," in *2022 IEEE International Conference of Electron Devices Society Kolkata Chapter (EDKCON)*, Kolkata, India, pp. 1–6, 2022, https://doi.org/10.1109/EDKCON56221.2022.10032940.
25. A. Goel, S. Rewari, S. Verma, and R. S. Gupta, "Dielectric modulated triple metal gate all around MOSFET (TMGAA) for DNA biomolecule detection," in *Proceedings of International Conference on 2018 IEEE Electron Device Kolkata Conference, EDKCON 2018*, pp. 337–340, 2018.
26. A. Goel, S. Rewari, S. Verma, S. S. Deswal, and R. S. Gupta, "Dielectric modulated junctionless biotube FET (DM-JL-BT-FET) biosensor," *IEEE Sensors Journal*, vol. 21, no. 15, pp. 16731–16743, 2021.
27. K. N. Singh, and P. K. Dutta, "Analytical modeling of underlap graded channel field effect transistor as a label-free biosensor," *Superlattices and Microstructures.*, vol. 155, 2021, https://doi.org/10.1016/j.spmi.2021.106897.
28. S. Kumar, B. Singh, and Y. Singh, "Analytical model of dielectric modulated trench double gate junctionless FET for biosensing applications," *IEEE Sensors Journal*, vol. 21, no. 7, pp. 8896–8902, 2021, https://doi.org/10.1109/JSEN.2021.3056385.
29. A. Kaul, S. Rewari, and D. Nand, "Hetero-dielectric macaroni channel cylindrical gate all around field effect transistor (HD-MC CGAA FET) for reduced gate leakage analog applications," *Microsystem Technologies.*, vol. 30, no. 5, pp. 599–611, 2024, https://doi.org/10.1007/s00542-023-05577-9.
30. S. Yadav, and S. Rewari, "Dual metal dual layer GAA NW–FET (DMDL–GAA–NW–FET) biosensor for label free SARS-CoV-2 detection," *Microsystem Technologies* pp. 1–19, 2023, https://doi.org/10.1007/s00542-023-05560-4.
31. S. Yadav, and S. Rewari, "Trench gate JAM dielectric modulated nanowire FET (TG-JAM-DM-NWFET) biosensor," in *Proceedings of the 2022 IEEE International Conference of Electron Devices Society Chapter, EDKCON 2022*, vol. 1, pp. 23–28, 2022.
32. S. Yadav, and S. Rewari, "Analytical modeling and numerical simulation of graded JAM Split Gate-All-Around (GJAM-SGAA) Bio-FET for label free Avian Influenza antibody and DNA detection," *Microelectronics*, vol. 142, p. 106011, 2023.
33. A. Das, S. Rewari, B. K. Kanaujia, S. S. Deswal, and R. S. Gupta, "Numerical modeling of a dielectric modulated surrounding-triple-gate germanium-source MOSFET (DM-ST-GGS-MOSFET)-based biosensor," *Journal of Computational Electronics.*, vol. 22, no. 2, pp. 742–759, 2023, https://doi.org/10.1007/s10825-023-02008-w.
34. A. Das, S. Rewari, B. K. Kanaujia, S. S. Deswal, and R. S. Gupta, "Physics based numerical model of a nanoscale dielectric modulated step graded germanium source biotube FET sensor: Modelling and simulation," *Physica Scripta*, vol. 98, no. 11, p. 115013, 2023, https://doi.org/10.1088/1402-4896/acf4c9.

35. S. Yadav, A. Das, and S. Rewari, "Dielectrically-modulated GANFET biosensor for label-free detection of DNA and Avian influenza virus: Proposal and modeling," *ECS Journal of Solid State Science and Technology* vol. 13, no. 4, p. 047001, 2024, https://doi.org/10.1149/2162-8777/ad3364.
36. S. Rashid, F. Bashir, F. A. Khanday, and M. R. Beigh, "Double gate 6H - silicon carbide Schottky barrier FET as dielectrically modulated label free biosensor," *Silicon*, vol. 15, pp. 3387–3398, 2023, https://doi.org/10.1007/s12633-022-02273-7.
37. Ashima, V. Dhandapani, and B. Raj, "Design and performance assessment of graded channel gate-all-around silicon nanowire FET for biosensing applications," *Silicon*, vol. 15, no. 8, pp. 3535–3542, 2023, https://doi.org/10.1007/s12633-022-02272-8.
38. N. Shaw, and B. Mukhopadhyay, "Modeling and performance analysis of a split-gate T-shape channel DM DPDG-TFET label-free biosensor," *IEEE Sensors Journal*, vol. 23, no. 2, pp. 1206–1213, 2023, https://doi.org/10.1109/JSEN.2022.3224036.
39. B. Cao *et al.*, "How to optimize materials and devices via design of experiments and machine learning: Demonstration using organic photovoltaics," *ACS Nano*, vol. 12, pp. 7434–7444, 2018.
40. N. C. Frey, D. Akinwande, D. Jariwala, and V. B. Shenoy, "Machine learning-enabled design of point defects in 2D materials for quantum and neuromorphic information processing," *ACS Nano*, vol. 14, no. 10, pp. 13406–13417, 2020.
41. L. Bassman *et al.*, "Active learning for accelerated design of layered materials," *npj Computational Materials.*, vol. 4, no. 1, p. 71, 2018, http://dx.doi.org/10.1038/s41524-018-0129-0.
42. M. Lee, "Machine learning for understanding the relationship between the charge transport mobility and electronic energy levels for n-type organic field-effect transistors," *Advanced Electronic Materials.*, vol. 5, no. 12, p. 1900573, 2019.
43. J. Song *et al.*, "A sequential multidimensional analysis algorithm for aptamer identification based on structure analysis and machine learning," *Analytical Chemistry Journal*, vol. 92, pp. 3307–3314, 2020.

Index

E

F

M

N

O

P

R

S

T

U

V

Y